A Reference Guide to Electrochemical Materials Science

A Reference Guide to Electrochemical Materials Science

Edited by **Bruce Hopkin**

C WILLFORD PRESS

New York

Published by Willford Press,
118-35 Queens Blvd., Suite 400,
Forest Hills, NY 11375, USA
www.willfordpress.com

A Reference Guide to Electrochemical Materials Science
Edited by Bruce Hopkin

International Standard Book Number: 978-1-68285-017-6 (Hardback)

The publisher's policy is to use permanent paper from mills that operate a sustainable forestry policy. Furthermore, the publisher ensures that the text paper and cover boards used have met acceptable environmental accreditation standards.

Trademark Notice: Registered trademark of products or corporate names are used only for explanation and identification without intent to infringe.

Printed in the United States of America.

Contents

Preface

Materials science is an interdisciplinary field of study integrating principles from physics, chemistry, engineering, etc. This book is compiled in such a manner, that it will provide in-depth knowledge about the theory and applications of electrochemical materials sciences. Also included in this book are detailed explanations of the various concepts such as electrochemical catalysis, electrochemical synthesis, corrosion, etc. It presents researches and studies performed by experts across the globe. Those in search of information to further their knowledge will be greatly assisted by this book.

The researches compiled throughout the book are authentic and of high quality, combining several disciplines and from very diverse regions from around the world. Drawing on the contributions of many researchers from diverse countries, the book's objective is to provide the readers with the latest achievements in the area of research. This book will surely be a source of knowledge to all interested and researching the field.

In the end, I would like to express my deep sense of gratitude to all the authors for meeting the set deadlines in completing and submitting their research chapters. I would also like to thank the publisher for the support offered to us throughout the course of the book. Finally, I extend my sincere thanks to my family for being a constant source of inspiration and encouragement.

Editor

Hardness and electrochemical behavior of ceramic coatings on Inconel

C. SUJAYA✉, H. D. SHASHIKALA, G. UMESH, A. C. HEGDE*

Department of Physics, National Institute of Technology Karnataka, Surathkal, Srinivasnagar 575025, Karnataka, India
* Department of Chemistry, National Institute of Technology Karnataka, Surathkal, Srinivasnagar 575025, Karnataka, India

✉Corresponding Author: E-mail: sujayachendel@gmail.com

Abstract

Thin films of ceramic materials like alumina and silicon carbide are deposited on Inconel substrate by pulsed laser deposition technique using Q-switched Nd: YAG laser. Deposited films are characterized using UV-visible spectrophotometry and X-ray diffraction. Composite microhardness of ceramic coated Inconel system is measured using Knoop indenter and its film hardness is separated using a mathematical model based on area-law of mixture. It is then compared with values obtained using nanoindentation method. Film hardness of the ceramic coating is found to be high compared to the substrates. Corrosion behavior of substrates after ceramic coating is studied in 3.5% NaCl solution by potentiodynamic polarization and electrochemical impedance spectroscopy measurements. The Nyquist and the Bode plots obtained from the EIS data are fitted by appropriate equivalent circuits. The pore resistance, the charge transfer resistance, the coating capacitance and the double layer capacitance of the coatings are obtained from the equivalent circuit. Experimental results show an increase in corrosion resistance of Inconel after ceramic coating. Alumina coated Inconel showed higher corrosion resistance than silicon carbide coated Inconel. After the corrosion testing, the surface topography of the uncoated and the coated systems are examined by scanning electron microscopy.

Keywords

Microhardness; Alumina; Silicon carbide; Pulsed laser deposition; Potentiodynamic polarization; Electrochemical impedance spectroscopy.

Introduction

Nickel-base superalloy Inconel (NiCrFe alloy) is commonly used in aircraft gas engines, rocket engines, nuclear-power plants, petrochemical equipment and offshore industries [1,2]. In these applications, components must have reliable chemical and mechanical stability at high operating temperatures. Pitting corrosion of Inconel has also been reported [3] when used in components operating in corrosive medium. To overcome these problems, the surface of the alloy is usually coated with ceramic materials. Coatings of ceramic materials are especially suitable for the protection against wear and corrosion [4,5,6]. Alumina (Al_2O_3) and silicon carbide (SiC) are the ceramic materials widely used for surface coatings. These coatings can be produced by many chemical and physical vapor deposition techniques. One of them, Pulsed Laser Deposition (PLD), is considered as flexible, simple and easily controllable method for producing thin films [7]. Indentation microhardness is a reliable test method for the evaluation of the mechanical property of the coatings. The Knoop indenter is used for microhardness measurements. The Knoop indentation depth is shallower [approximately 1/30 the long diagonal] than the Vickers indentation depth [approximately 1/7 the average diagonal] when the same load is applied on the material. Hence, the Knoop indentation is better suited for testing of thin coatings. Hardness of thin films has combined effect of the substrate and the film that leads to the so-called "composite hardness". The composite hardness includes the component of the substrate hardness depending on the relative depth of penetration of the indenter and mechanical properties of both, the film and the substrate. Film hardness is separated from the composite hardness based on the area law of mixtures. Film hardness can also be determined by nanoindentation technique in which load-displacement curve is obtained by depth sensing indentation technique.

In the present study, efforts were made to get adhesive films of Al_2O_3 and SiC using PLD technique on the fine polished Inconel substrate at 450^0C. Usually thicker ceramic films are deposited at temperatures higher than 600^0C. In the present investigation adhesive films of lesser thickness (about 0.5 µm) were deposited at lower temperature (450 °C) and appreciable increase in microhardness and corrosion resistance after coating was observed. Processing parameters, like laser fluence, substrate target distance, and substrate temperature during the deposition are standardized using multiple trials. Composite microhardness of Al_2O_3 and SiC coatings are measured using Knoop indenter. Its film hardness is separated using a model based on Johnson and Hogmark [8], after including the effect of indentation size effect (ISE) [9], and compared with the values obtained using nanoindentation technique. A corrosion behavior of the coated substrate is studied in 3.5% NaCl solution by potentiodynamic polarization and Electrochemical Impedance Spectroscopy (EIS) measurements.

Experimental

The substrate used is Inconel 601. The substrates are sequentially polished with SiC waterproof abrasive paper from 320 grit to 1500 grit size and with Al_2O_3 suspension of 1, 0.5, 0.3 µm size to an average roughness (R_a) of 40 nm. The polished surfaces are cleaned with acetone in ultrasonic bath for 20 minutes before the deposition of coating on the substrate.

A Q-switched Nd: YAG laser (Laser spectra) of wavelength 1064 nm and frequency 10 Hz was used for deposition. Laser beam of energy of 135 mJ focused to 0.1 cm diameter using a spherical lens of focal length 50 cm was impinged onto the rotating Al_2O_3 target. Inconel substrates of size 2×2 cm^2 were used for coating. Sintered Al_2O_3 and SiC pellets of thickness 0.4 cm and diameter 4 cm were used as targets. The rotation speed of 40 rotations per minute was used to avoid the crater formation. The incident angle between laser beam and target surface was kept at 45°. To obtain a uniform film thickness, the substrate was placed at 8 cm distance from the target surface.

The ablation was carried out in vacuum under the pressure of 10^{-3} Pa, and at a substrate temperature of 450 °C. Figure 1 shows the schematic representation of the PLD system used in the present study.

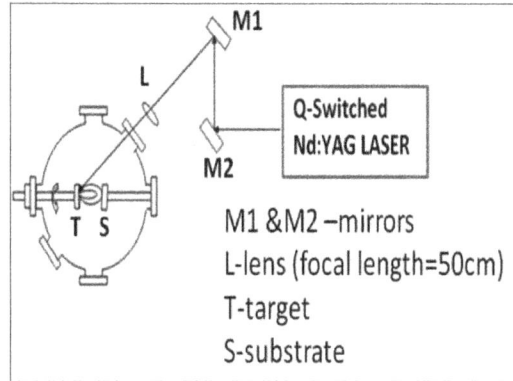

Figure 1. *Schematic representation of the PLD system.*

Measurement techniques

The average surface roughness was measured using 3-D optical surface profilometer (Vecco). Stylus profilometer (Taylor Hubson) was used to measure the film thickness. Optical absorption measurements were performed on Al_2O_3 and SiC coated glass substrate at room temperature using UV-Visible spectrophotometer (Ocean optics). XRD analysis (JEOL) was carried out by utilizing CuK_α radiation on ceramic coated substrates.

Microhardness tester (Clemex) with Knoop indenter was used to measure the composite hardness. Loads of 25, 50, 100, 200, 300 and 500 g were employed at a dwell time of 10 seconds. At each load 10-12 indentations were made on the sample surface. The average microhardness at each load was calculated.

In the Knoop hardness measurement, the Knoop hardness is obtained from the following relation

$$HK = 14229 \frac{P}{L^2} \tag{1}$$

where *HK* is Knoop hardness in GPa, 14229 is a constant dependent on the indenter geometry. *P* is the applied load in N, *L* is the length of long diagonal of the resultant rhombic impression in μm.

The microhardness measurement of the coating is the composite hardness of the film-substrate composite system. To separate the composite hardness into its constituents, a model based on area *i.e.,* area "law-of mixture" approach is used. It was originally proposed by Jonsson and Hogmark [8] for Vickers indentation and later it is adapted to the geometrical configuration of the Knoop indenter [10]. The composite hardness, H_c of the film-substrate system is expressed as

$$H_c = \frac{A_f}{A}H_f + \frac{A_s}{A}H_s \tag{2}$$

where A is the contact area; H is the hardness; the subscripts c, f and s are related to the composite, film and the substrate, respectively and $A = A_f + A_s$. By means of simple geometrical relations depending on the Knoop indentation, equation for A_f / A can be written as

$$\frac{A_f}{A} = \frac{2ct}{L} - \frac{c^2 t^2}{L^2} \tag{3}$$

where L is the length of the large diagonal, t is the thickness of the film, $c = 2.908$ for brittle film on the soft surface or $c = 5.538$ for soft film on the soft surface [11]. Usually, measured hardness varies with load due to ISE which has not been taken into account in Jonsson and Hogmark model. Jonsson and Hogmark model is improved by incorporating the ISE taking into account a linear relation between hardness and the reciprocal indentation depth [12]. To obtain the true hardness of the film, the hardness variation with the applied load is written as

$$H_f = H_{fo} + \frac{B_f}{L} \tag{4}$$

$$H_s = H_{so} + \frac{B_s}{L} \tag{5}$$

where H_{fo} and H_{so} are the absolute hardness of the film and the substrate respectively, B_f and B_s are constants and L is the indentation diagonal length. Substituting equation (3), (4) and (5) in (2) and neglecting second order $1/L$ terms, the composite hardness becomes

$$H_c = H_{so} + \frac{B_c}{L} \tag{6}$$

where $B_c = B_s + 2ct(H_{fo} - H_{so})$. To evaluate H_{so} and B_s values, hardness measurements on substrate was performed. The experimental plot H_s versus $1/L$ is approximated well by a linear regression. From the intercept and slopes of these plots, H_{so} and B_s of Inconel are obtained as 3.6 GPa and 37.98 GPa m^{-1}, respectively. The experimental data on the composite hardness of the film-substrate system, H_c is plotted against $1/L$ and the intrinsic hardness of the film (H_{fo}) was calculated from the slope of the regression line using equation (6). Film hardness obtained from the above model was compared with the film hardness obtained using nano-indentation method.

Generally, it is not possible to measure the properties of a thin film independent of the substrate, using conventional testing equipment, unless coating is very thick. To determine the film hardness, the indentation depth should be 10% of the film thickness. In such cases nanoindenter was used. Common feature of this method is that it continuously monitors the load and the displacement as the indentation is produced. The feature of a continuous depth and load recording allows thin film properties to be obtained directly from the data without the need to measure indentation diagonals. The data were analyzed with the Oliver and Pharr method [13]. Hysitron Tribo indenter was used for the nanoindentation.

The corrosion behavior of the coated and uncoated substrates were studied using Tafel polarization studies by immersing samples in 3.5% NaCl solution in open air and at room temperature, using potentiostat/galvanostat from Gamry Instruments (PCI 4G750-47065). All electrochemical measurements were performed using conventional three electrode cell, using a platinum plate as an auxiliary electrode, saturated calomel electrode (SCE) as a reference electrode. The exposed area working electrode to the corrosive medium was 1cm^2. The sample was cleaned in distilled water before loading it to the sample holder. The sample was placed in such a way that Luggin capillary of the reference electrode was close to the working electrode and this arrangement was used for all the tests. The open circuit potential (E_{ocp}) or the steady state potential is maintained to study the potentiodynamic polarization. After achieving the stable E_{ocp}, the upper and the lower potential limits of the linear potential sweep were set at ± 250 mV vs E_{ocp}. The sweep rate was 1 mV s^{-1}. The Tafel plot was drawn using the obtained electrochemical data. The corrosion potential (E_{corr}) and the corrosion current density (i_{corr}) are deduced from the Tafel plot (log$I$ vs. E plot). The corrosion current density is obtained using the Stern–Geary equation [14]:

$$i_{corr} = \frac{B}{R_p} \tag{7}$$

where $B = b_a\,b_c\,/2.303(b_a + b_c)$ and where B is called Stern–Geary constant, b_a and b_c are Tafel slopes for the anodic and cathodic reactions respectively and R_p is the polarization resistance expressed in $\Omega\ cm^2$. The corrosion rate (CR) in micrometer per year (y) was calculated from the Faraday's law [15] using the following equation

$$CR = \frac{3.27 i_{corr} EW}{d} \tag{8}$$

where EW is the equivalent weight of the testing material in grams and d is the density of the testing sample in g cm^{-3}. By substituting the value of i_{corr}, CR was calculated.

Impedance measurements were conducted using the frequency response analyzer (FRA). The spectrum was recorded in the frequency range 10 mHz - 300 kHz. The applied alternating potential had the amplitude of 10 mV. After each experiment, the impedance data were displayed as Nyquist and Bode plots. A circuit description code (CDC) is assigned for the acquired data and the acquired data are curve fitted and analyzed using ZSimpWin software.

Results and Discussion

Optical characterization of the coatings

Adhesive coatings of Al_2O_3 and SiC were obtained on Inconel substrate by using the standardized parameters described in the previous section. The optical band gap energy and Urbach energy calculations for SiC and Al_2O_3 coated on the glass substrate were carried out using optical absorption data. The Tauc's plot is used to obtain the optical band gap energy (E_g) [16]:

$$\alpha h\nu = \mathrm{const}(h\nu - E_g)^p \tag{9}$$

where E_g is the optical band gap energy and $h\nu$ is the photon energy of the incident radiation, the exponent p in the equation is 2 for the indirect optical transition and $p=1/2$ for the direct optical transition. The energy band gap E_g of SiC and Al_2O_3 film is obtained by extrapolating linear region of the curve obtained by plotting $(\alpha h\nu)^{1/2}$ as a function of $h\nu$ in the case of SiC (indirect band gap) and $(\alpha h\nu)^2$ versus $h\nu$ in the case of Al_2O_3 (direct band gap) as shown in Figures 2 and 3. The E_g values are found to be 3.32 eV for Al_2O_3 and 2.07 eV for SiC film. These values match with the reported values of 3.2 eV for Al_2O_3 [17] and 2-2.9 eV for SiC [18]. Generally, amorphous films have low band gap energies compared to crystalline films.

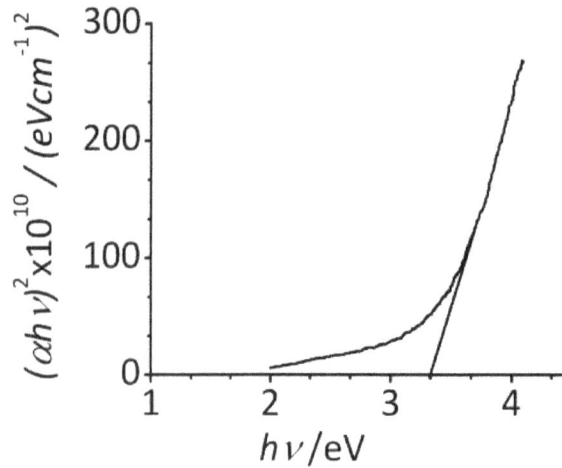

Figure 2. *Absorption coefficient versus energy for the Al_2O_3 film.*

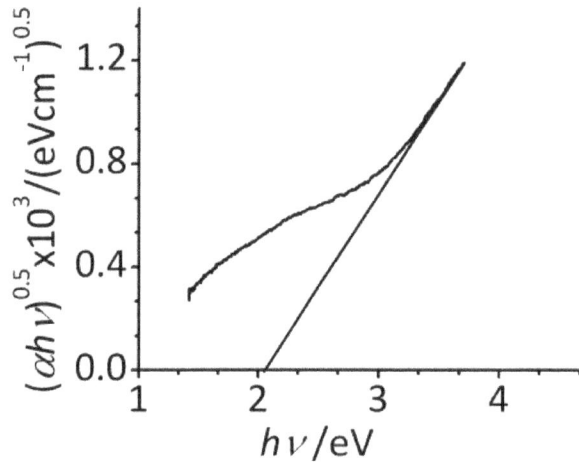

Figure 3. *Absorption coefficient versus energy for the SiC film.*

The obtained coated films are amorphous in nature. This was confirmed by optical absorption data. At longer wavelength in the spectrum, there is an absorption tail described by Urbach exponential law [19], which describes the amorphous nature of the film. The Urbach energy is related to the absorption coefficient as

$$\alpha h v = \alpha_o e^{\frac{h v}{E_u}} \tag{10}$$

where α_o is a constant. The Urbach energy E_u is the measure of a disorder in the system.

Figures 4 and 5 correspond to the plot of $\ln\alpha$ as a function of hv. The value of Urbach energy was calculated by fitting the linear region of Figures 4 and 5 applying the method of least squares to the straight line equation ($\ln\alpha = hv / E_u + $ constant) and was found to be 1.49 eV for Al_2O_3 and 1.39 eV for SiC and film. Urbach energy E_u is the measure of structural disorder in the system. The value of Urbach energy obtained for SiC and Al_2O_3 confirms that the obtained films are amorphous in nature, which is confirmed also by the absence of sharp XRD peaks.

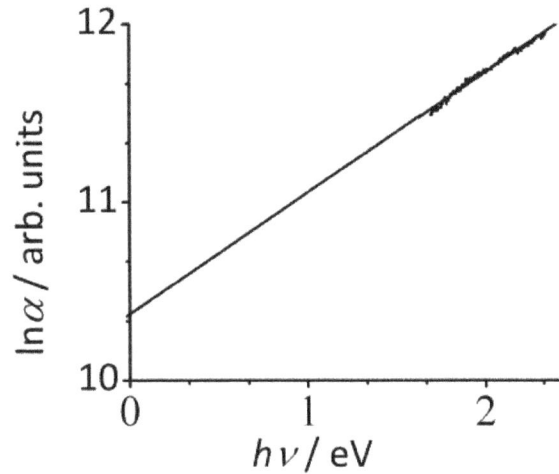

Figure 4. *lnα as a function of energy (hν) for Al$_2$O$_3$ film.*

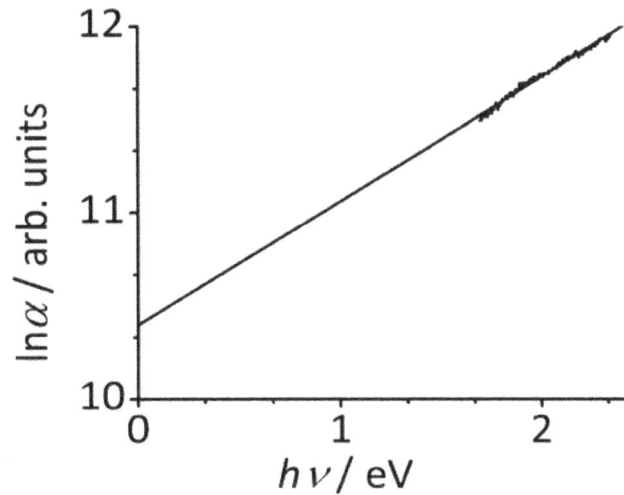

Figure 5. *lnα as a function of energy (hν) for SiC film.*

Hardness measurements

Figures 6a and 6b show the dependence of the measured composite film-substrate hardness on the inverse of indentation diagonal for Al$_2$O$_3$ and SiC coatings on Inconel. A least-square fit to the plots of the equation 6 results in the slope $B_c = [B_s + 2ct(H_{fo}-H_{so})]$. The value of c is chosen depending on the film whether it is brittle or soft, since the nature of the films depends on their bulk material. Its value is 2.908 for brittle film and 5.538 for soft film [11,20]. To verify whether the film is brittle or soft we observed the nature of indentation on the surface of the coating. If the film is brittle it accommodates the indenter by crack formation. If it is a soft film, it plastically deforms to match the shape of the indenter. In the present study, there was no cracking of the film observed in Al$_2$O$_3$ and SiC deposited substrates even at high load (10 N) which is shown in Figure 7. A soft nature of the film is due to the amorphous structure of the coated film [21,22]. Since the deposition was carried out at low temperature (450 °C), deposited ceramic films were not transformed into crystalline form as confirmed by optical absorption and XRD data.

From the above reasoning, assuming the film to be soft, the c value is selected as $c = 5.538$. Using the value of B_s and H_{so} of substrate, the intrinsic hardness of the films was calculated using the equation 6. The slope of the straight line obtained by plotting H_c versus $1/L$ (Figure 6) was used to calculate the intrinsic film hardness H_{fo}. The intrinsic hardness of the coated films is listed in Table 1. The film hardness gives of higher value than that of the substrate.

Nano-indentation measurements were performed on the surfaces of both, uncoated Inconel and with Al_2O_3 and SiC, coatings in order to determine the hardness of coating. Figure 8 compares the indentation load-depth curve for Inconel substrate and Al_2O_3 film and SiC film of thickness 0.5 µm on Inconel. The indentation is carried out at the peak load of 1000 µN. The difference in hardness of the two samples is clearly evident from the difference in the depth obtained at the maximum load. The uncoated sample exhibits a larger indentation depth at the maximum load. In order to find the intrinsic hardness of the coated film an indentation depth should be 0.1 of the film thickness. The load was selected in such a way that its indentation depth was less than 0.05 µm on the coated substrate. Table 2 compares the hardness value obtained using mathematical model, and the nano-indentation method and it shows that the two values are in a good agreement.

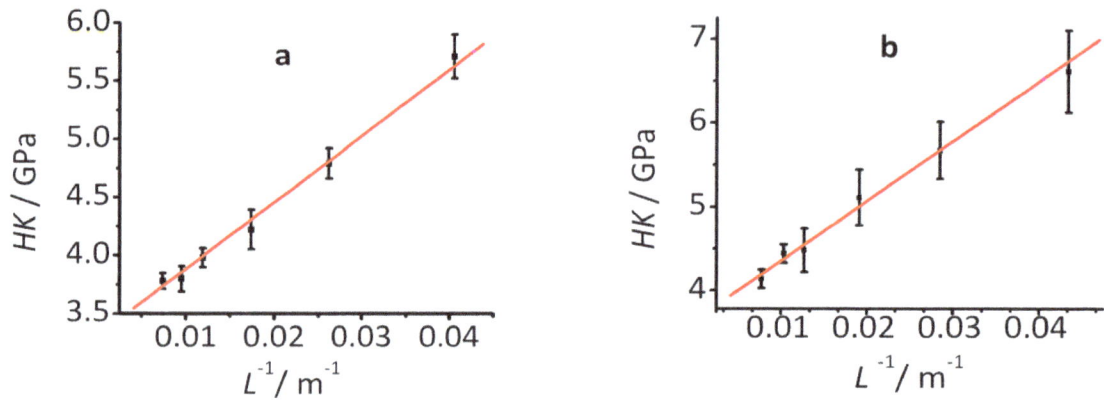

Figure 6. *Knoop hardness variation with 1/L on a) Al_2O_3 and b) SiC coated film on Inconel.*

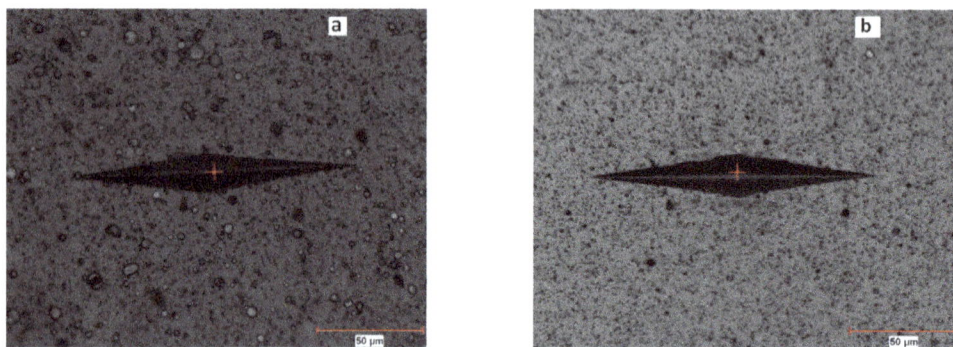

Figure 7. *Knoop indentation on a) Al_2O_3 b) SiC coated film on Inconel substrate at 10N load.*

Table 1. *Hardness data of coatings on Inconel*

Coating material	t / µm	B_s / GPa m^{-1}	B_c / GPa m^{-1}	$2c$	H_{so} / GPa	H_{fo} / GPa
Al_2O_3	0.5	37.98	56.90	11.076	3.61	7.0
SiC	0.5	37.98	64.74	11.076	3.61	8.6

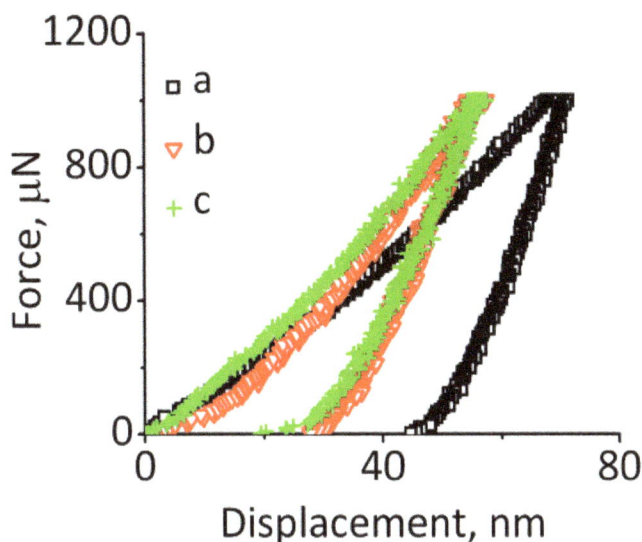

Figure 8. *Load variations with nanoindentation depth*
a) substrate Inconel b) Al$_2$O$_3$ and c) SiC coating on Inconel.

Table 2. *Comparison of hardness values obtained using a model and nanoindentation method.*

Material	Surface roughness, nm	Thickness of the film, μm	Hardness, GPa	
			From Model	Nanoindentation
Al$_2$O$_3$ on Inconel	42.46	0.5	7.0	7.74
SiC on Inconel	40.46	0.5	8.6	8.65

Potentiodynamic polarization measurements

The Tafel plots obtained for Inconel, Inconel with Al$_2$O$_3$ and Inconel with SiC coatings of thickness (t) 0.5 μm in 3.5 % NaCl solution at 30 °C are shown in Figure 9. The corrosion potential (E_{corr}), the corrosion current density (i_{corr}) and the corrosion rate are deduced from the Tafel (log$i$ vs. E) plots. The values are listed in Table 3. For the Inconel substrate the i_{corr} is approximately 1.007 nA cm^{-2} and it decreases to 0.235 nA cm^{-2} for alumina coatings and to 0.850 nA cm^{-2} for SiC coatings. Corrosion rate is directly proportional to $i_{corr.}$. It is established that CR reduced drastically after ceramic coating. Alumina coated substrates exhibit excellent corrosion resistance. Corrosion rate after Al$_2$O$_3$ coating reduced to 1/4 of that of Inconel, whereas SiC coated samples reduced to 1/2 of Inconel. The improvement in the corrosion resistance of the substrate after ceramic coating is due to the fact that ceramic coating passivates the surface of the substrate and prevents the corrosion attack by electrolyte. Generally, in physical vapor deposited coatings, electrolytes enter into the substrate through the defects like pores and micro cracks and cause the corrosion of the substrate [23]. The protective coatings must fulfill the two main requirements to achieve the remarkable effect on corrosion resistance: a strong adhesion to the substrate and a low density of pores and cracks. The absence of pores and micro cracks completes surface coverage and good adhesion of the ceramic film coating to the substrate obtained by PLD technique enhances the corrosion resistance of the substrate. Enhanced corrosion rate of coatings on Inconel in the present study indicates that the coated films are having negligible concentration of pores and cracks.

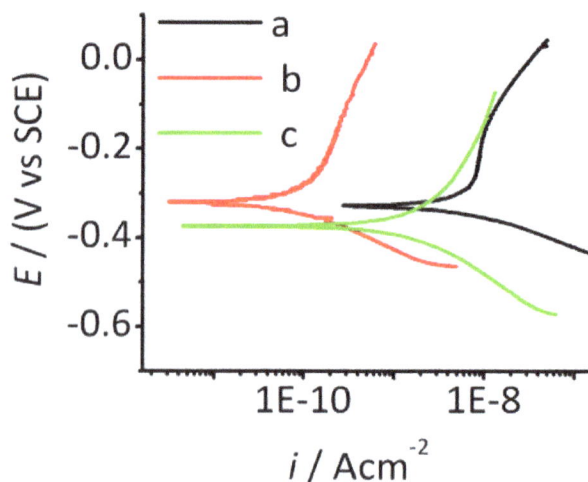

Figure 9. Potentiodynamic polarization curves of
a) Inconel, b) Al$_2$O$_3$ coated Inconel and c) SiC coated Inconel.

Table 3. *Potentiodynamic polarization measurements.*

	Inconel	SiC	Al$_2$O$_3$
b_a / mV dec^{-1}	110.9	125.9	62.59
b_c / mV dec^{-1}	44.9	78	165.5
i_{corr} / nA cm^{-2}	3.69	1.007	0.235
E_{corr} / mV	-328.0	-372.0	-321.0
CR /μm year^{-1}	44.00	29.95	12.78
R_p / MΩ cm^{-2}	3.76	20.8	83.9

Electrochemical impedance spectroscopy studies

Figures 10, 11 and 12 show the impedance response of Inconel, Inconel with Al$_2$O$_3$ and Inconel with SiC coatings as Nyquist and Bode plots. The EIS data obtained from these plots are listed in Table 4. The Nyquist plots show unfinished semicircles what is attributed to the charge transfer controlled reactions [24]. Bode plots in the form of phase angle vs. log ω showed that the phase angles remain almost constant between higher and lower frequencies.

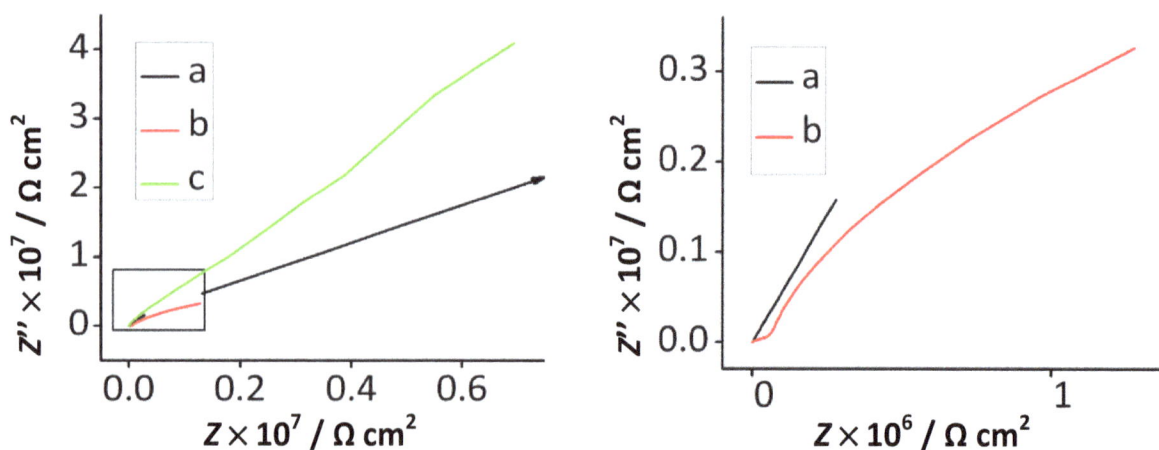

Figure 10. *Nyquist plots of: a) Inconel, b) SiC and c) Al$_2$O$_3$ coatings.*

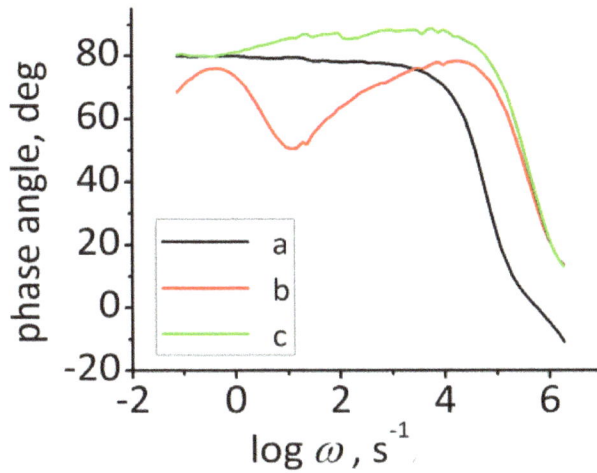

Figure 11. Bode plots (phase angle vs. logω) of a) Inconel, b) SiC and c) Al$_2$O$_3$ coatings.

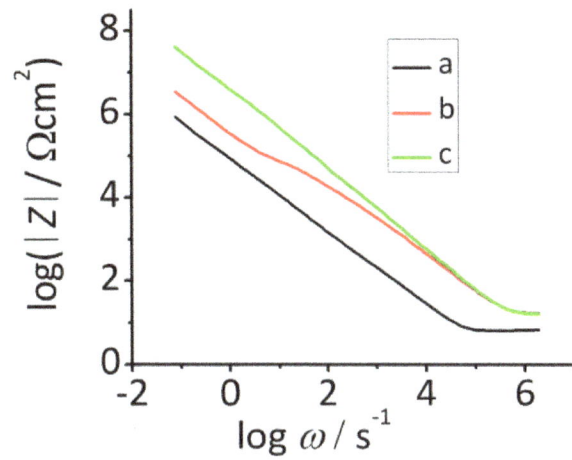

Figure 12. Bode plots (log⎸Z⎸vs. logω) of a) Inconel, b) SiC and c) Al$_2$O$_3$ coatings.

The simplified equivalent circuits are shown in Figure 13. The values of solution resistance (R_s), charge-transfer resistance (R_{ct}), pore resistance (R_p), total capacitance (Q) of the constant phase element (CPE) are listed in Table 4. The R_{ct} increases in the following order: Inconel substrate < < SiC coated Inconel < Al$_2$O$_3$ coated Inconel, which shows that the alumina coated Inconel has the highest corrosion resistance. The values of R_{ct} are strongly dependent on the passive film characteristics and are an indication for corrosion of materials. This supports the results obtained using potentiodynamic polarization measurements. log⎸Z⎸ vs. logω plots of Fig. 12 show that the absolute impedance increased in the same order as mentioned above. The equivalent circuit for Inconel is shown in Fig. 13(a). It consists of a double layer capacitance, which is parallel to the charge transfer resistance, both of which are in series with the solution resistance between the working electrode (WE) and the tip of the Luggin capillary. The double layer capacitance provides information about the polarity and the amount of charge at the substrate/electrolyte interface. For better quality of fitted results, pure capacitance is replaced with constant phase element (CPE) that accounts for deviation from ideal dielectric behavior. Its admittance, $Y=Z^{-1}$, is defined as:

$$Y(\omega) = Q(j\omega)^n \qquad (11)$$

where Q is an adjustable parameter used in the non-linear least squares fitting and n is also an adjustable parameter that always lies between 0.5 and 1. The value of n is obtained from the slope of log⎸Z⎸ versus log ω plot (Figure 12). The phase angle (θ) can vary between -90° (for a perfect capacitor (n = 1)) and 0 (for a perfect resistor (n = 0)). The CDC for the equivalent circuit proposed for Inconel is R(Q)R. When the sample is immersed in the electrolyte, the defects in the coating provide the direct diffusion path for the corrosive media. In this process the galvanic corrosion cells are formed and the localized corrosion dominates the corrosion process. In such cases, electrochemical interface can be divided into two sub-interfaces: electrolyte/coating and electrolyte/substrate. The proposed equivalent circuit for such a system is shown in Figure 13(b). The parameters in the equivalent circuit R_p and Q_c are related to the properties of the coating and the electrolyte/coating interface reactions. R_{ct} and Q_{dl} are related to the charge-transfer reaction at the electrolyte/substrate interface. The CDC of the proposed equivalent circuit for the coated sample is R(Q[R(QR)]). From the EIS data given in Table 4, it is evident that Q_c decreased from SiC coating to alumina coating, indicating that SiC coatings contain relatively more pores and less dense microstructure as compared to alumina coatings.

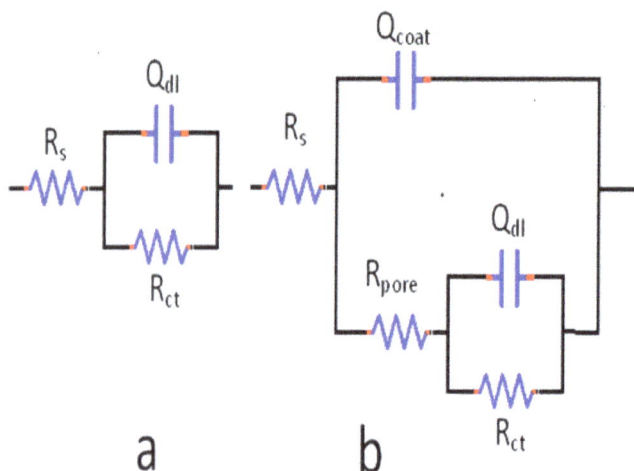

Figure 13. *a) equivalent circuit to fit the electrochemical impedance data of the Inconel substrate and b) equivalent circuit to fit the electrochemical impedance data of the coated substrate.*

Scanning Electron Microscopy (SEM) Studies

After the corrosion testing, the surface morphology of the uncoated and the coated system are studied by SEM. The corrosion pits were observed on the uncoated Inconel after the corrosion test (Figure 14a). Corrosion pits are not found in Al_2O_3 and SiC coated samples, (Figures 14b and 14c)). Pits are formed as a result of corrosion attack on the uncoated surface. The improvement in the corrosion resistance is due to adherent thin Al_2O_3 film coating deposited on the surface of the substrate which will not allow the direct contact of the electrolyte with the substrate. EDAX analysis indicates the presence of sodium (Na) and chlorine (Cl) in the regions where the pits are formed. Similar conclusion can be drawn in corrosion behavior of SiC coated Inconel.

Table 4. *EIS data obtained by equivalent circuit simulation of SiC and Al_2O_3 coatings.*

Sample	R_s Ω cm^{-2}	Q_{dl} $\Omega^{-1}s^n$	n_{dl}	R_{ct} MΩcm^{-2}	Q_{coat} $\Omega^{-1}s^n$	n_c	R_{pore} MΩcm^{-2}
Inconel	15.46	2.07	0.8	0.89			
Inconel/SiC	14.66	0.212	0.89	21.07	0.73	0.87	7.66
Inconel/Al_2O_3	15.82	0.16	0.99	108	0.085	0.74	18.6

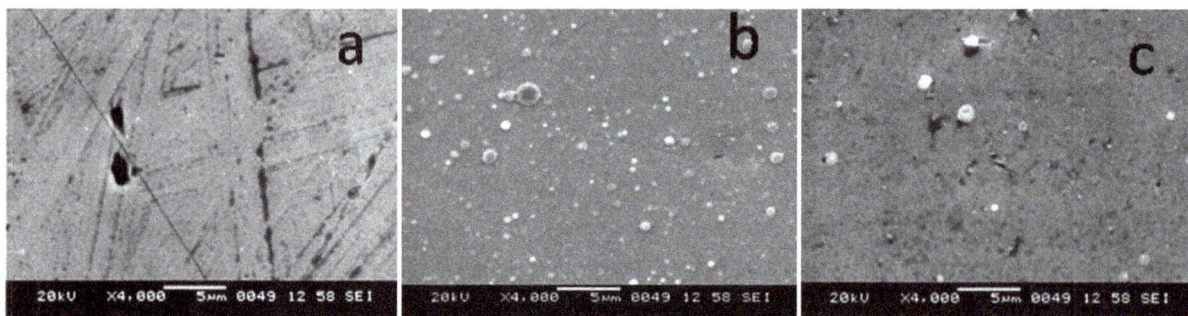

Figure 14. *Surface morphology of the a) Inconel substrate b) alumina c) SiC coated substrate surface.*

Conclusions

Adhesive thin films of Al_2O_3 and SiC have been deposited on Inconel substrate by PLD technique using Nd: YAG laser of wavelength 1064 nm. Knoop microhardness of the film-substrate system gives the composite microhardness. Film hardness can be separated from the composite hardness using a model based on area law of mixtures. It is nearly equal to the film hardness measured using nanoindentation technique. The potentiodynamic polarization and the EIS measurements showed that Al_2O_3 and SiC coatings exhibit better corrosion resistance as compared to that of substrate.

Acknowledgements: *The authors would like to thank Indian Space Research Organization (ISRO) for financial assistance in carrying out this research and to Vikram Sarabhai Space Centre (VSSC), Trivandrum for supplying substrate materials.*

References

[1]   S.H. Jeong, C.W. Cho, Z. Lee, *Tribol. Int.* **38** (2005) 283–288

[2]   C.J. Wang, S.M. Chen, *Surf. Coat. Tech.* **201** (2006) 3862–3866

[3]   Y.I. Kim, H.S. Chung, W.W. Kim, J.S. Kim, W.J. Lee, *Surf. Coat. Tech.* **80** (1996) 113-116

[4]   C. Cibert, H. Hidalgo, C. Champeaux, P. Tristant, C. Tixier, J. Desmaison, A. Catherinot, *Thin Solid Films* **516** (2008) 1290-1296

[5]   M. Zaytouni, J.P. Rivière, *Wear* **197** (1996) 56-62

[6]   Y. Li, J. Yao, Y. Liu, *Surf. Coat. Tech.* **172** (2003) 57-64

[7]   T.J. Zhu, L. Lu, M. O. Lai, *Appl. Phys. A* **81(4)** (2005) 701-714

[8]   B. Jönsson, S. Hogmark, *Thin Solid Films* **114** (1984) 257-269

[9]   G. Guillemot, A. Iost, D. Chicot, *Thin Solid Films* **518** (2010) 2097-2101

[10]   A. Iost, *Scripta Mater.* **39** (1998) 231-238

[11]   F. Torregrosa, L. Barrallier, L. Roux, *Thin Solid Films* **266** (1995) 245-253

[12]   A. Iost, R. Bigot, *Surf. Coat. Tech.* **80** (1996) 117-120

[13]   W.C. Oliver, G.M. Pharr, *J. Mater. Res.* **7** (1992) 1564-1583

[14]   V.S. Sastri, *Corrosion Inhibitors–Principles and Applications*, John Willy and Sons, Chichester, 1998.

[15]   S.W. Dean, in *Electrochemical Techniques for Corrosion*, R. Baboian, Ed., National Association of Corrosion Engineers; Houston, TX, USA, 1977, p. 52–60

[16]   J. Tauc, *Optical Properties of Solids*, F. Abels Ed., North-Holland, Amsterdam, 1972, p. 277

[17]   V. Rose, V. Podgursky, I. Costina, R. Franchy, *Surf. Sci.* **541** (2003) 128–136

[18]   Y.J. Park, Y.W. Park, J.S. Chun, *Thin Solid Films* **166** (1988) 367-374

[19]   F. Urbatch, *Phys. Rev.* **92** (1953) 1324

[20]   D. Ferro, S.M. Barinov, J.V. Rau, A. Latini, R. Scandurra, B. Brunetti, *Surf. Coat.Tech.* **200** (2006) 4701-4707

[21]   J. Esteve, A. Lousa, E. Martinez, H. Huck, E.B. Halac, M. Reinoso, *Diam. Relat. Mater.***10** (2001) 1053-1057

[22]   M. Sridharan, M. Sillassen , J. Bottiger , J. Chevallier , H. Birkedal, *Surf. Coat. Techn.* **202** (2007) 920–924

[23]   H.C. Barshilia, K. Yogesh, K.S. Rajam, *Vacuum* **83** (2009) 427-434

[24]   L. Liu, Y. Li, F. Wang, *Electrochim.Acta* **52** (2007) 7193-7202

2

Optimization of parameters for dye removal by electro-oxidation using Taguchi Design

Mani Nandhini, Balasubramanian Suchithra,

Ramanujam Saravanathamizhan✉ and Dhakshinamoorthy Gnana Prakash

Department of Chemical Engineering, SSN College of Engineering, Kalavakkam, Chennai 603110, India

✉Corresponding author

Abstract

The aim of the present investigation is to treat the dye house effluent using electro-oxidation and to analyse the result using Taguchi method. L16 orthogonal array was applied as an experimental design to analyse the results and to determine optimum conditions for acid fast red dye removal from aqueous solution. Various operating parameters were selected to study the electro-oxidation for the colour removal of the effluent. The operating parameter such as dye concentration, reaction time, solution pH and current density were studied and the significance of the variables was analysed using Taguchi method. Taguchi method is suitable for the experimental design and for the optimization of process variables for the dye removal.

Keywords
Electro-oxidation; Taguchi design; colour removal; acid fast red; optimization

Introduction

Effluents discharged from textile industries have high intensity of colour, which leads to pollution. Highly coloured wastewater can be treated by different methods such as biological treatment, chemical coagulation, activated carbon adsorption, ultrafiltration, ozonation, wet oxidation, photocatalysis, electrochemical methods *etc.* [1-5]. Among these methods the electrochemical treatment has been receiving greater attention in recent years due to its unique features such as versatility, energy efficiency, automation and cost effectiveness [6]. The electrochemical technique offers high removal efficiencies and the main reagent is the electron called 'clean reagent' which degrades all the organics present in the effluent without generating any secondary pollutant or by-product/sludge.

Industrial wastewaters have been treated by electro-oxidation techniques and the operating parameters have been optimized by different techniques. The widely used technique is response surface methodology for the experimental design and process optimization. Taguchi method is another method in the experimental design, proposed by Genichi Taguchi, contains system design, parameter design, and tolerance design. Taguchi method was effectively used to improve the product or process effectiveness by using a loss function to attain the product quality in terms of the parameter design [7]. Taguchi is the preferable technique among statistical experimental design methods since it uses a special design of an orthogonal array to study the effective parameters with a minimum number of experiments. This method helps researchers to determine the possible combinations of factors and identify the best combination. However, in industrial settings, it is extremely costly to run a number of experiments to test all combinations. The parameter design using Taguchi method minimizes the time and experimental runs. In the design, 'signal' and 'noise' (S/N) represent the desirable and undesirable values for the output characteristics, respectively and the ratio is a measure of the quality characteristic deviating from the desired value. Further an analysis of variance (ANOVA) is used to determine the significant parameters.

Recently Taguchi's designs have been applied to various chemical and environmental engineering studies for the experimental design and for the optimization of the process variables. Asghari et al. [8] used Taguchi method to determine the optimum conditions for methylene blue dye removal from aqueous solutions using electrocoagulation. The authors found that the amount of electrolyte was the most significant parameter for the colour removal. Srivastava et al. [9] used Taguchi method to determine the optimum conditions for the orange-G dye removal from aqueous solution by electrocoagulation using iron plate electrodes. Author also applied this methodology to optimize the process variables for the multi-component adsorption of metal ions onto bagasse fly ash and rice husk ash [10]. Kaminari et al. [11] used Taguchi method to determine the optimum condition of process variables and find the influencing parameters for the recovery of heavy metals from acidified aqueous solutions using electrochemical reactor. Kim et al. [12] optimized the experimental process variable using Taguchi technique for the nano-sized silver particles production by chemical reduction method. In another study Mohammadi et al. [13] used Taguchi method to determine the optimal experimental conditions for the separation of copper ions from a solution. Moghaddam et al. [14] used Taguchi method to design the experimental runs and optimize the parameters for the ammonium carbonate leaching of nonsulphide zinc ores. Maria et al. [15] designed the experimental runs and optimized the process parameters using Taguchi method for the adsorption of acid orange 7 dyes on guava seed. In that way Taguchi method was used to study the electro oxidation of dye removal.

The objective of the present study is to treat the acid fast red dye using electro-oxidation. The effect of experimental parameters such as initial dye concentration, reaction time, solution pH and current density on colour removal was investigated using an L16 orthogonal array. The Taguchi experimental design has been used to determine the optimum conditions for the maximum colour removal of the dye from aqueous solutions.

Materials and Methods

All the chemicals used in this study were AR grade (Merck). Acid fast red was prepared by dissolving definite quantity of dye in the distilled water. To increase the conductivity of the solution of 1 mg L^{-1} NaCl was used as supporting electrolyte for all experimental runs. The initial solution pH was adjusted by 0.1 M HCl or 0.1 M sodium hydroxide solution. Experiments were repeated twice to minimize the experimental error.

Experimental setup

The experimental setup of the batch reactor is shown in the Figure 1. The volume of the reactor was 250 ml and electrodes used were fixed inside the reactor with 1 cm space between them. Stainless steel sheet cathode and mesh type Ruthenium oxide coated Titanium anode were used. The void fraction of the mesh type anode accounts 20 % by area, which resulted in an effective anode area of 28 cm^2 (7×5 cm). The electrodes were connected to a 5 A, 10 V DC regulated power supply, through an ammeter and a voltmeter. The solution was constantly stirred at 200 rpm using a magnetic stirrer in order to maintain a uniform concentration. DC power supply was given to the electrodes according to the required current density and the experiments were carried out under constant current conditions. The samples were analysed for the colour removal using UV-Vis spectrophotometer (Jasco, V-570). The percentage colour removal was calculated by:

$$\text{Colour removal, \%} = \frac{Abs_i - Abs_t}{Abs_i} \times 100 \tag{1}$$

where *Abs*$_i$ and *Abs*$_t$ are absorbance of initial and at time *t* at the corresponding wavelength λ_{max}.

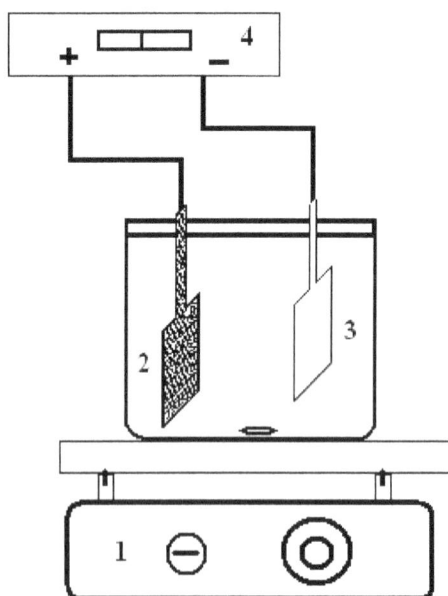

Figure 1. *Schematic diagram of the experimental setup*
1. Magnetic stirrer 2. Anode 3. Cathode 4. DC Power supply

Taguchi design

The following procedure was adopted for the parameter design.
1. Planning of experiment
    i.   Determine the experimental responses of the process.
    ii.   Determine the levels of each variable.
    iii. Select a suitable orthogonal array table. The selection based on the number of variables and number of levels.
    iv.   Transform the data from the experiments into a proper S/N ratio.
2. Implementing the experiment, based on design table.

3. Analyzing and examining the result
    i.   ANOVA analysis to determine the significant parameters in the process.
    ii.   Draw the main effect plot, S/N ratio plot, mean plot to analysis the optimal level of the control variables.

The factors and levels chosen for the present experiment are shown in the Table 1. L16 orthogonal array design was selected for the four variables with four different levels for the each factor. Table 2 shows the Taguchi design for the electro oxidation of acid fast red dye. Each row represents one experimental run. Based on Taguchi design the experiments were carried out and the percentage colour removal was observed as response. The proposed design was an orthogonal array, for which each pair of the columns had all the possible combinations of levels. The *S/N* ratio characteristics can be divided into three categories when the characteristic is continuous:

(i)   Nominal is the best characteristic

$$\frac{S}{N} = 10 \log \frac{\bar{y}}{s_y^2}$$
(2)

Table 1. Variables and their values corresponding to their levels investigated in the experiments

	Variables	Level			
		1	2	3	4
A	Dye concentration, mg l^{-1}	25	50	75	100
B	Time, min	15	30	45	60
C	pH	2	5	7	10
D	Current density, mA cm^{-2}	5	7.5	10	12.5

Table 2. Experimental variables, their levels and results of conducted experiments corresponding to L16 experimental plan

S. No	Levels				Run 1	Run2
	A	B	C	D	Colour removal, %	
1	1	1	1	1	78.50	78.40
2	1	2	2	2	81.82	81.20
3	1	3	3	3	86.42	86.40
4	1	4	4	4	89.75	90.00
5	2	1	2	3	70.08	69.98
6	2	2	1	4	84.41	84.15
7	2	3	4	1	77.20	78.01
8	2	4	3	2	91.53	91.50
9	3	1	3	4	71.55	71.54
10	3	2	4	3	68.21	68.12
11	3	3	1	2	86.78	86.78
12	3	4	2	1	83.44	83.58
13	4	1	4	2	52.05	52.19
14	4	2	3	1	56.70	57.25
15	4	3	2	4	91.66	91.89
16	4	4	1	3	96.32	96.80

(ii) Smaller the better characteristics

$$\frac{S}{N} = -10\log\frac{1}{n}\sum y^2 \qquad\qquad (3)$$

(iii) Larger the better characteristics

$$\frac{S}{N} = -10\log\frac{1}{n}\sum\frac{1}{y^2} \qquad\qquad (4)$$

Where, $\bar{y}$ is the average of observed data, s_y^2 the variance of y, n is the number of observations, and y the observed data. For each type of the characteristics, with the above S/N ratio transformation, the higher the S/N ratio the better is the result. The experimental data were analysed using MINITAB 14 (PA, USA) [16].

Result and discussion

Main effect plot

Main effect plot for the percentage colour removal using electro oxidation of acid fast red dye is shown in the Figure 2. The plot is used to visualize the relationship between the variables and output response. The effect of initial dye concentration on colour removal is shown by the factor 'A'. The percentage colour removal of the dye increases with decrease in dye concentration. This is due to the fact that at higher initial dye concentrations, the intermediate products formed due to the degradation of dyes increase the resistance of current flow by blocking the electrode active sites, and thus, decrease colour removal. The effect of electrolysis time on colour removal is shown in the figure by a factor 'B'. The percentage colour removal depends on the electrolysis time. When time increases the generation of OCl⁻ radical increases due to electro-oxidation which results increase the percentage colour removal. The effect of pH on the mean colour removal is represented by the factor 'C'. As it is observed from the figure, colour removal increases with the decrease of pH. This is due the fact that the hydroxyl radical generation is high at acidic pH which results to increase the rate of colour removal. The effect of current density on the removal of dye is shown by the factor 'D'. The increase of current density increases the OCl⁻ generation hence the percentage colour removal increases. High concentrations of chloride ions and salts in water can improve the performance and effectiveness of the electro-oxidation process. Various levels (1, 2, 3, 4) of the operating parameter (A, B, C, D) and their mean colour removal is shown in the Table 3. It is observed form the table, Level '1' shows highest colour removal of 84.06 %, 84.37 % for the initial dye concentration and pH, respectively. Level '4' shows highest colour removal of 90.37 %, 86.52 % for electrolysis time and current, respectively.

Table 3. *Mean colour removal for electro oxidation of acid fast red*

Level	Mean colour removal, %			
	A	B	C	D
1	84.06	68.04	86.52	74.14
2	80.86	72.74	81.71	77.98
3	77.50	85.64	76.62	80.29
4	74.36	90.37	71.94	84.37

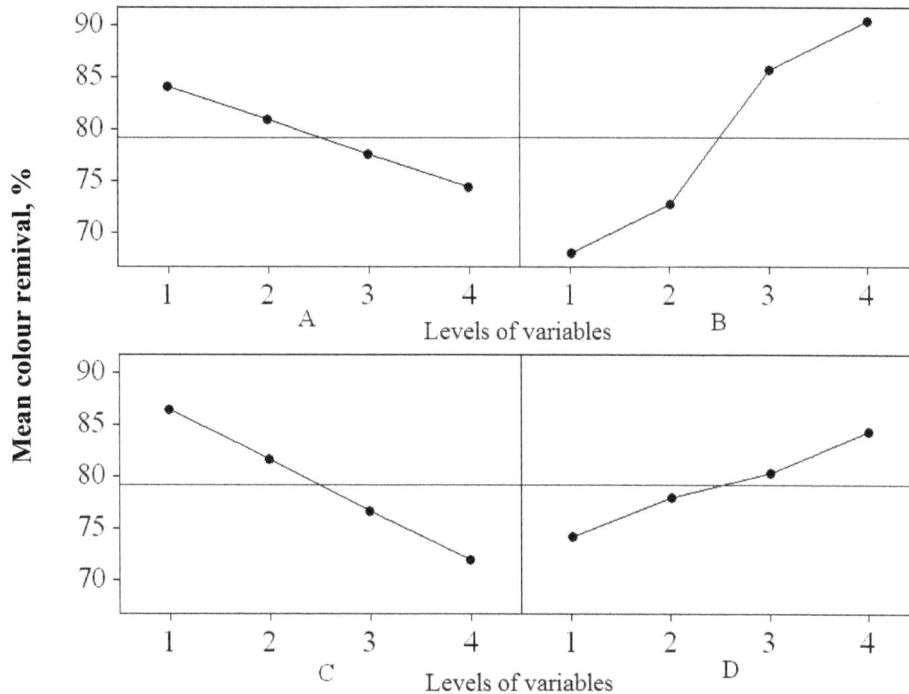

Figure 2. *Main effect plot for the percentage colour removal of acid fast red*
A: Dye concentration parameter; B: Time parameter; C: pH parameter; D: Current density parameter

Signal to noise (S/N) ratio

Taguchi method was used to identify the optimal conditions and most influencing parameters on colour removal. In the Taguchi method, the terms 'signal' to 'noise' ratio represent the desirable and undesirable values for the output response, respectively. The *S/N* ratios are different according to the type of output response. In the present case, larger *S/N* ratio is better for high colour removal. Figure 3 shows the *S/N* ratio of dye removal using electro-oxidation.

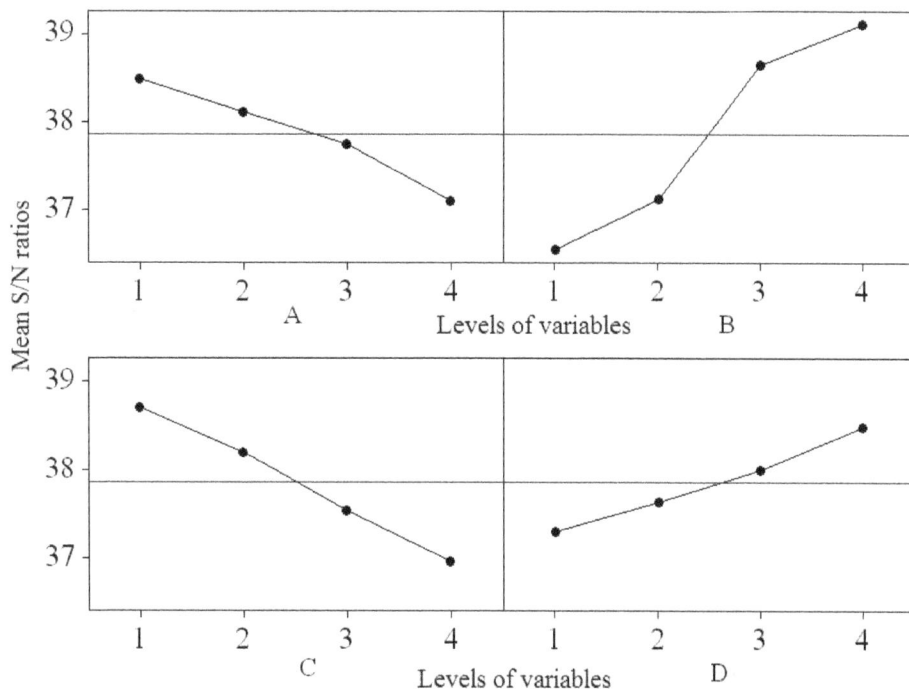

Figure 3. *S/N ratio plot for the percentage colour removal of acid fast red .*
A: Dye concentration parameter; B: Time parameter; C: pH parameter; D: Current density parameter

It can be noticed that at higher *S/N* ratio better level for colour removal was achieved. It is observed that factor 'A', initial dye concentration, and factor 'C', pH, are required in a lower level and electrolysis time 'B' and applied current density 'D' are required at higher level for the maximum colour removal.

Optimization of process parameters based on S/N ratio

The process parameters were optimized based on S/N ratio. Lager the S/N ratio higher the percentage colour removal and vice versa. The values of the S/N ratios for the operating parameters are shown in the Table 4. It is observed from the Table that the S/N ratio of level '1' 38.48, 38.72 shows the higher value for the factor 'A' and 'C', respectively. It shows that level '1' gives the higher colour removal. The larger S/N ratios of factor 'B' and 'D' for the level '4' are 39.11, 38.48, respectively. It shows that level '4' gives the higher colour removal.

Table 4. *S/N ratio for electro oxidation of acid fast red*

Level	S/N ratio of electro oxidation			
	A	B	C	D
1	38.48	36.56	38.72	37.31
2	38.11	37.13	38.20	37.64
3	37.74	38.64	37.54	38.00
4	37.10	39.11	36.97	38.48

Analysis of variance (ANOVA)

Analysis of variance (ANOVA) was performed to see whether the process parameters were statistically significant or not. The F-test is a tool to check which process parameters have a significant effect on the colour removal. The P value less than 0.05 shows that the parameter is significant. ANOVA table for the colour removal of the dye is shown in the Table 5. It is observed form the table the most influential factor was electrolysis time because the P value is 0.049 with the corresponding sum of the square is higher compared to other variable in the Table. Then the less significant variables are pH, current density and initial dye concentration.

Table 5. *ANOVA table for the electro oxidation of acid fast red*

Source	Degree of freedom	Sum of the squares	Mean of squares	F	P
A	3	210.9	70.29	1.50	0.374
B	3	1330.5	443.48	9.46	0.049
C	3	476.8	158.94	3.39	0.171
D	3	220.2	73.40	1.56	0.361
Residual Error	**3**	**140.7**	**46.90**		
Total	**15**	**2379**			

Conclusion

Taguchi experimental design was used to determine the optimum operating conditions of the dye removal from aqueous solutions using electro oxidation. The significant variables were identified for the colour removal process. Optimum levels for operating parameters can be simul-taneously identified with the Taguchi method. The advantage of the Taguchi method is the reduction in time and minimization of the number of experimental runs. For the present study level '1' is best for initial dye concentration (25 g L^{-1}) and initial pH (2), level '4' is best for electrolysis time (60 min) and current density (12.5 mA cm^{-2}) for the highest colour removal. It can be concluded that Taguchi method is suitable for the experimental design and to optimize the process variable for the colour removal of the dye effluent.

Acknowledgement: The authors are grateful to the SSN Trust for the financial support of this work.

References

[1]   M. C. Gutierrez, M. Crespi, *Journal of the Society of Dyers and Colorists* **115** (1999) 342–345.

[2]   M. Panizza, G. Cerisola, *Chemical Reviews* **109** (2009) 6541–6569.

[3]   C. A. Martinez-Huitle, E. Brillas, *Applied Catalysis B: Environmental* **87** (2009) 105-145.

[4]   L. C. Davies, I. S. Pedro, J. M. Novais, S. Martins-Dias, *Water Research* **40** (2006) 2055 - 2063.

[5]   G. Chen, *Separation & Purification Technology* **38** (2004) 11–41.

[6]   F. C. Walsh, *Pure and Applied Chemistry* **73** (2001) 1819–1837.

[7]   J. A. Ghani, I. A. Choudhury, H. H. Hassan, *Journal of Materials Processing Technology* **145** (2004) 84–92.

[8]   A. Asghari, M. Kamalabadi, H. Farzinia, *Chemical and Biochemical Engineering Quarterly* **26** (2012) 145–154.

[9]   V. C. Srivastava, I. D. Mall, I. M. Mishra, *Chemical Engineering Journal* **140** (2008) 136–144

[10]   V. C. Srivastava, I. D. Mall, I. M. Mishra, *Industrial & Engineering Chemistry Research* **46** (2007) 5697-5706.

[11]   N. M. S. Kaminari, D. R. Schultz, M. J. J. S. Ponte, H. A. Ponte, C. E. B. Marino, A. C. Neto, *Chemical Engineering Journal* **126** (2007)139-146.

[12]   K. D. Kim, D. N. Han, H. T. Kim, *Chemical Engineering Journal* **104** (2004) 55-61.

[13]   T. Mohammadi, A. Moheb, M. Sadrzadeh, A. Razmi, *Desalination* **169** (2004) 21-31.

[14]   J. Moghaddam, R. Sarraf-Mamoory, M. Abdollahy, Y. Yamini, *Separation & Purification Technology* **51** (2006)157-164.

[15]   M. P. Elizalde-Gonzalez, V. Hernandez-Montoya, *Journal of Hazardous Materials* **168** (2009) 515–522.

[16]   K. Ravikumar, S. Ramalingam, S. Krishnan, K. Balu, *Dyes and Pigments* **70** (2006) 18-26.

3

A comparison of pitting susceptibility of Q235 and HRB335 carbon steels used for reinforced concrete

GUOLIANG ZHAN, JIANCHENG LUO*, SHAOFEI ZHAO*, WEISHAN LI*⊠

New Century Concrete of Guangdong Foundation Co., Ltd, Guangzhou 510660, China

*School of Chemistry and Environment, South China Normal University, Guangzhou 510006, China

⊠Corresponding Author

Abstract

The phase structure and the pitting susceptibility of two carbon steels, Q235 and HRB335, used for reinforced concrete, are investigated by phase observation, polarization curve measurements, electrochemical impedance spectroscopy, and Mott-Schottky analysis. It is found that Q235 is ferrite and HRB335 is pearlite. Q235 is more susceptible to chloride ions leading to pitting than HRB335. The polarization curves show that the breakdown potential of the passive film in saturated $Ca(OH)_2$ solution containing 0.4 M NaCl is 0 V for Q235 and 0.34 V for HRB335. The Mott-Schottky analyses show that passive films formed on Q235 and HRB335 in saturated $Ca(OH)_2$ solution containing chloride ions behave like an n-type semiconductor. The passive film formed on Q235 has a higher donor density, which explains why Q235 is more susceptible to pitting than HRB335.

Keywords

Pitting susceptibility; Carbon steel; Phase structure; Polarization curve; Electrochemical impedance spectroscopy; Mott-Schottky.

Introduction

Reinforced concrete is widely used as building material because the corrosion resistance of the embedded carbon steel for the reinforcement plays a significant role in the life of reinforced concrete. In a high quality concrete, the embedded carbon steel prevents corrosion by forming a passive film on a steel surface, which slows down the access of oxygen, moisture, and various aggressive species to the interface between steel and concrete [1]. Among all the aggressive species, chloride ions exhibit the strongest attack on passive film. Chloride ions may be introduced to the concrete from raw material, such as water and sand, or they can penetrate from the outside in highway viaducts where de-icing salts are used as well as in marine structures [2]. In practice, the corrosive attack due to chloride penetration usually leads to pitting corrosion, which causes catastrophe because its initiation and propagation is difficult to predict.

The pitting susceptibility of carbon steel is related to the microstructure and composition of the steel and inhibitors used for the formation of passive films [3-12]. Various carbon steels usually are combined for use because of the need for strength and tenacity. For example, two kinds of carbon steels, Q235 and HRB335, are usually used together in reinforced concrete structures to improve the strength of concrete. It is necessary to understand the pitting susceptibility of various kinds of carbon steel for their successful application in reinforced concrete. The aim of this work is to understand the difference in pitting susceptibility between Q235 and HRB335.

Experimental

Electrodes and solutions

The composition of Q235 and HRB335 is given in Table 1. The steel specimens (ϕ 0.8 cm × 0.5 cm) were embedded in epoxy resin, with a test area of 0.5 cm^2. Prior to each measurement, the working surface of specimens was polished with different SiC_2 abrasive papers and Al_2O_3 powder of 0.05 μm, then degreased with ethanol and rinsed with de-ionized water successively.

The saturated $Ca(OH)_2$ solution was prepared as simulated concrete pore solution (SPS) with de-ionized water. NaCl was added to form the solution containing chloride ions. All the chemicals used were of analytical grade.

Table 1. Composition of Q235 and HRB335, wt. %

Samples	C	Mn	P	S	Si	Cr	Ni	Al	As
Q235	0.15	0.326	0.039	0.03	0.115	0.024	0.032	0.127	0.028
HRB335	0.26	0.698	0.046	0.05	0.324	0.033	0.039	0.016	0.026

Phase-structure observation

Optical microscope (ECLIPSE 50iPOL, Nikon Corporation) was used to observe the microstructure of Q235 and HRB335. Before the observation, the specimens were etched with 0.2 % Nital for 10s, degreased with ethanol and rinsed with de-ionized water.

Electrochemical measurement

The electrochemical measurements were carried out with PGSTAT-30 (Autolab, Eco Chemise B. V. Company). A classical three electrodes electrochemical cell was used. A platinum sheet with a geometric area of 1 cm^2 was used as the counter electrode and a saturated calomel electrode (SCE) was used as the reference electrode. All the potentials in this paper are versus the SCE.

The electrolyte was deareated with nitrogen for 30 min before each measurement, and all the electrochemical measurements were carried out at ambient temperature without stirring. The potential scanning rate used in the polarization curve measurements was 0.5 mV s^{-1}. The impedance measurements were carried out in frequency range from 100 kHz to 0.01 Hz. In the Mott-Schottky measurement, the frequency used was 1 kHz and the potential step was 5 mV. Prior to each measurement, the working electrode was firstly kept at -1.15 V for 30 min, then at 0 V for 60 min to form a passive film, and finally it was stabilized at open circuit potential (OCP) for 30 min.

Results and Discussion

Phase structure

Fig. 1 presents the phase structure of Q235 (A) and HRB335 (B). Both, Q235 and HRB335 have two phases, pearlite and ferrite [11,13]. However, the proportion of ferrite and pearlite is different in HRB335 and Q235. Ferrite phase prevails in Q235 while in HRB335 prevails mainly pearlite. The composition of two carbon steel is similar except for the contents of Mn and Si (Table 1), which might account for the difference in microstructure between two carbon steels. From the different microstructure of the two carbon steels, it can be expected that they exhibit different pitting susceptibility [3,9].

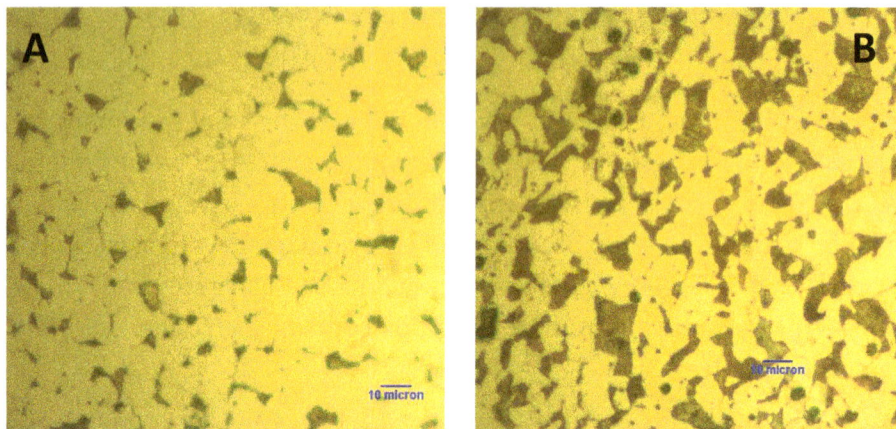

Figure 1. Phase structure of Q235 (A) and HRB335 (B)

Polarization curve

In order to understand the formation process of passive film and to determine the passive potential range, a potential scan ranging from -0.6 to 0.65 V is performed. Fig. 2 presents the formation of passive film on Q235 and HRB335 obtained in the simulated concrete pore solution (SPS). It can be seen from Fig. 2 that both electrodes, Q235 and HRB335, exhibit a similar formation process of passive film. A cathodic process takes place when the potential is scanned from point a (-0.6V) to b (-0.41 V) for Q235, and from point a (-0.6 V) to b' (-0.45 V) for HRB335. This cathodic process should be ascribed to the hydrogen evolution reaction, because the experiments were carried out under nitrogen atmosphere and there were no other reducible species in the solution. When potential is scanned from point b (-0.41 V) and b' (-0.45 V) toward positive potential, the current in both cases (Q235 and HRB335 electrodes) increases till the potential approaches point c (-0.18 V). From point c (-0.18 V) to point d (0.65 V) the electrodes reach passive region. Both electrodes, Q235 and HRB335, have similar passive current density of about 4 μA cm^{-2}. The sharp increase of the current density at point d (0.65 V) is ascribed to the O$_2$ evolution, which depends on pH of the solution [13]. The current plateau of the anodic polarization curve indicates the formation of passive film within the potential domain from -0.18 to 0.65 V for both carbon steel electrodes.

Fig. 3 shows the polarization curves of Q235 and HRB335 in the SPS containing 0.4 M chloride ions. Before the measurement, the two carbon steel electrodes were passivated in SPS, in absence of chloride ions, at the potential of 0 V for 60 min. In SPS containing 0.4 M chloride ions, the breakdown of passive films on both electrodes, Q235 and HRB335, occurs at the potential, which is indicated by a drastic increase of current. However, the breakdown potential values of passive films were different, *i.e.* about 0 V for Q235 and 0.34 V for HRB335. Therefore, the passive film formed on Q235 is more susceptible to chloride ions than the one formed on HRB335 electrode.

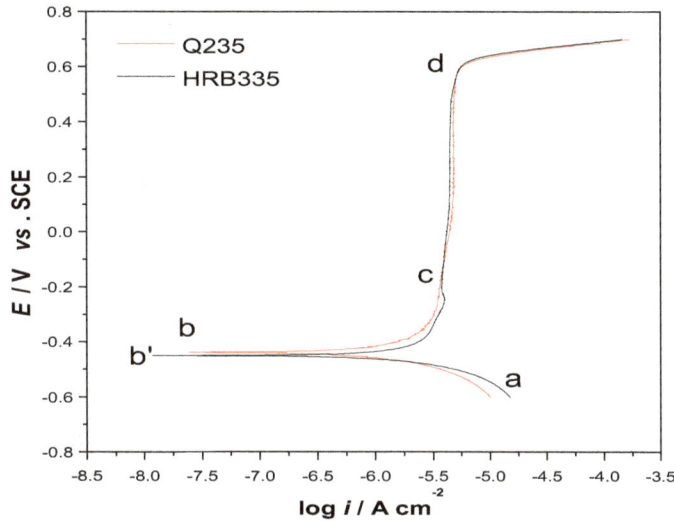

Figure 2. *Polarization curves of Q235 and HRB335 in SPS in absence of chloride ions (scan rate: 0.5 mV s^{-1}).*

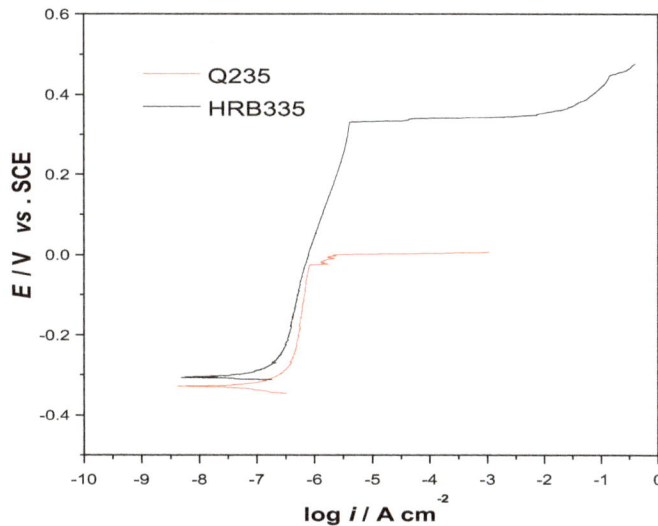

Figure 3. *Polarization curves of Q235 and HRB335 in SPS containing 0.4 M chloride ions (scan rate: 0.5 mV s^{-1}).*

Electrochemical impedance

Fig. 4 presents the Nyquist plots of passivated Q235 and HRB335 in the SPS containing 0 and 0.4 M chloride ions at open circuit potential. It can be seen from Fig. 4 that two electrodes have similar behavior which does not involve any diffusion process, although there is a significant difference in polarization impedance for the electrodes in the solution with and without chloride ions.

A passivated electrode can be modeled by the equivalent circuit of Fig. 5. In Fig. 5, R_S represents the solution resistance; R_f and Q are the resistance and the space charge layer capacitance of the passive film. The element Q is usually represented by the constant phase element (CPE) in which n is in the range between 0.5 and 1 due to the surface heterogeneity and surface roughness of the passivated electrodes [14]. The impedance of a CPE is given by

$$Z_{CPE} = \left[y^0 \left(jw \right)^n \right]^{-1}$$

(1)

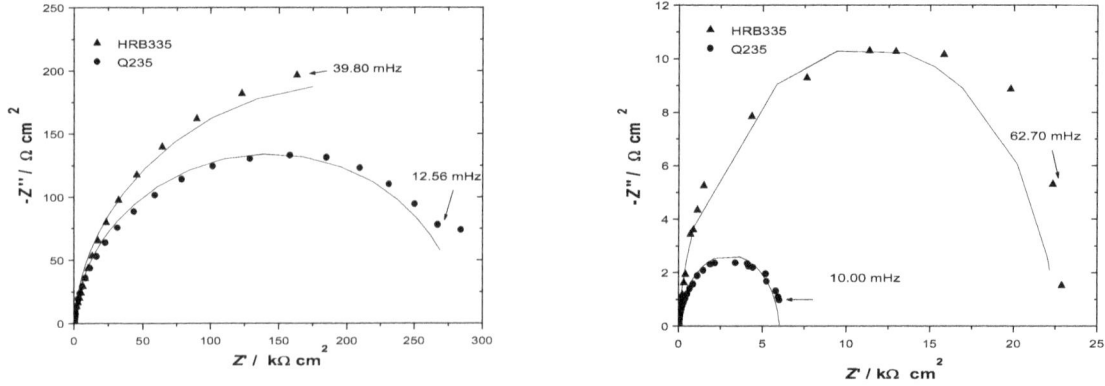

Figure 4. Nyquist plots of passivated Q235 and HRB335 electrodes in SPS: **(a)** without chloride ions; **(b)** containing 0.4 M chloride ions

The capacitance element Q (CPE) is pure capacitance when n = 1 and pure resistance when n = 0. The results obtained from fitting with Fig. 5 are shown by solid lines in Fig. 4 and by the data in Table 2. It can be seen from Table 2 that, in the solution without chloride ions, the film resistance (R_f) of Q235 is smaller than the one of HRB335, indicating that the passivated Q235 tends to react more easily than the passivated HRB335. When chloride ions are added in the solution, the resistance of the film decreases for both carbon steel electrodes.

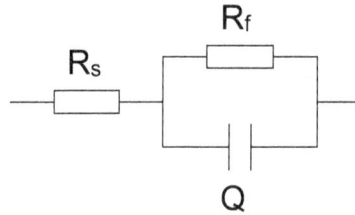

Figure 5. Equivalent circuit for a passivated electrode.

Mott-Schottky analysis

Passive films of most metals behave as semiconductor [15-18]. In the high frequency domain, the Mott-Shottly approach is a good tool for characterizing the semiconducting properties of a passive film [19,20]. When the frequency used for the impedance measurement is high enough, potential dependence of the capacitance of space-charge layer (C_{sc}) is expressed by Mott-Schottky relationship [5]:

For n-type semiconductor

$$\frac{1}{C_{sc}^2} = \frac{2}{e\varepsilon_r\varepsilon_0 N_D}\left(E - \phi_{fb} - \frac{kT}{e}\right) \tag{2}$$

For p-type semiconductor

$$\frac{1}{C_{sc}^2} = -\frac{2}{e\varepsilon_r\varepsilon_0 N_A}\left(E - \phi_{fb} - \frac{kT}{e}\right) \tag{3}$$

where e is electron charge (1.6×10^{-19} C), ε_r is dielectric constant, taken as 15.6 [17]. ε_0 is the vacuum permittivity (8.85×10^{-14} F cm^{-1}), N_D is donor density, N_A is acceptor density, E is the applied potential, φ_{fb} is flat-band potential, k is Boltzmann constant (1.38×10^{-23} J K^{-1}) and T is absolute temperature. N_D and N_A can be determined from the slope of the linear relationship between C_{sc}^{-2} and E, while ϕ_{fb} is obtained from the extrapolation to $C_{sc}^{-2} = 0$.

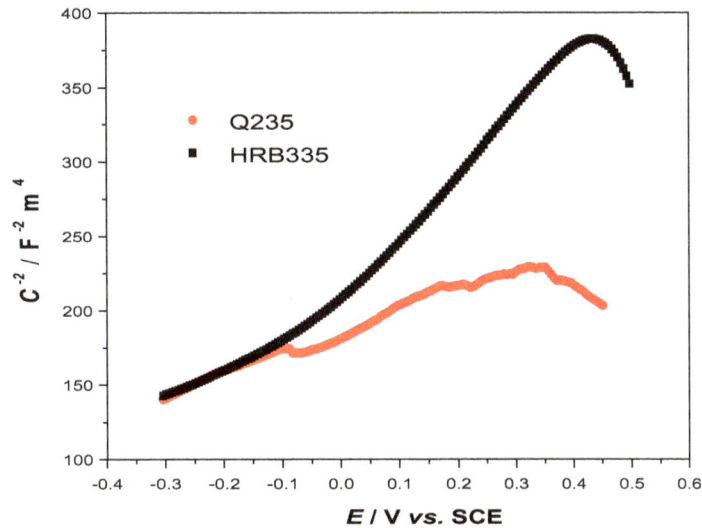

Figure 6. *Mott-Schottky plots of passivated Q235 and HRB335 in the SPS containing 0.4 M chloride ions.*

Fig. 6 shows Mott-Schottky plots of passivated Q235 and HRB335 electrodes in SPS containing 0.4 M chloride ions. It can be seen from Fig. 6 that, at the potentials lower than 0.35V, C_{sc}^{-2} increases with increasing the potential. There is a positive linear relationship between C_{sc}^{-2} and E in the potential range from 0 to 0.15 V for both electrodes, indicating that both passive films are n-type semiconductors [4]. The donor densities values calculated from the slope of the linear relationship between C_{sc}^{-2} and E in Fig. 6 are shown in Table 2. It is found that the donor densities values are very high, *i.e.* in the order of 10^{26} m^{-3}. This order is characteristic of heavily doped and disordered passive films, which was also found by Cheng and Luo (10^{26} m^{-3} $\leq N_D \leq 10^{27}$ m^{-3}) [21]. A donor density of a passive film determines its pitting susceptibility. The larger the donor density is, the more susceptible to pitting of the passive film is. The donor density of passive film formed on Q235 is 4.42×10^{26} m^{-3}, which is higher the one of HRB335 (2.29×10^{26} m^{-3}), confirming that Q235 is more susceptible to pitting than HRB335.

Table 2. *Fitting results of the experimental data in Fig. 4 and Fig. 6*

Samples	C_{NaCl} / M	R_f / kΩ cm^2	Y^0 /μS cm^{-2} s^n	n	N_D / 10^{26} m^{-3}
Q235	0	285.3	10.95	0.9607	-
Q235	0.4	6.0	18.42	0.9236	4.42
HRB335	0	402	10.81	0.9610	-
HRB335	0.4	22.5	12.55	0.9512	2.29

Conclusions

Based on the results from phase observation, potentiodynamic polarization, electrochemical impedance measurements and Mott-Schottky analysis of passive films of two carbon steel electrodes, it can be concluded that Q235 is more susceptible to pitting than HRB335. Q235 has a phase structure of ferrite, which tends to form a less stable passive film and thus is more susceptible to chloride ions than HRB335, whose phase structure is pearlite.

Acknowledgements*: This work was supported by the Natural Science Foundation of Guangdong Province (Grant No. 10351063101000001) and the joint project of Guangdong Province and*

Ministry of Education for the Cooperation among Industries, Universities and Institutes (Grant No. 2011B090400633).

References

[1]　A.A. Almusallam, *Constr. Build. Mater.* **15** (2001) 361-368

[2]　M. Ormellese, M. Berra, F. Bolzoni, T. Pastore, *Cement Concrete Res.* **36** (2006) 536-547

[3]　W.S. Li, J.L. Luo, *J. Mater. Sci. Lett.* **21** (2002) 1195-1198

[4]　W.S. Li, J.L. Luo, *Corros. Sci.* 44 (2002) 1695-1712

[5]　W.S. Li, N. Cui, J.L. Luo, *Electrochim. Acta* **49** (2004) 1663-1672

[6]　W.S. Li, S.Q. Cai, J.L. Luo, *J. Electrochem. Soc.* **151** (2004) B220-B226

[7]　J. Lu, W.S. Li, J.L. Luo, *Corros. Eng. Sci. Techn.* **43** (2008) 208-212

[8]　W.S. Li, J.L. Luo, *Int. J. Electrochem. Sci.* **2** (2007) 627-665

[9]　A. Pardo, M.C. Merino, A.E. Coy, R. Arrabal, F. Viejo, E. Matykina, *Corros. Sci.* **50** (2008) 823-834

[10]　E.E Abd El Aal, S. Abd El Wanees, A. Diab, S.M. Abd El Haleem, *Corros. Sci.* **51** (2009) 1611-1618

[11]　F. Zhang, J.S. Pan, C.J. Lin, *Corros. Sci.* **51** (2009) 2130-2138

[12]　L. Hamadou, A. Kadri, D. Boughrara, N. Benbrahim, J.P. Petit, *Appl. Surf. Sci.* **252** (2006) 4209-4217

[13]　D. Trejo, P.J. Monteiro, *Cement Concrete Res.* **35** (2005) 562-571

[14]　M. Cai, S.M. Park, *J. Electrochem. Soc.* **143** (1996) 3895-3902

[15]　A.M.P. Simoes, M.G.S. Ferrira, B. Rondot, M. CunhaBelo, *J. Electrochem. Soc.* **137** (1990) 82-87

[16]　G. Cooper, J.A. Turner, A.J. Nozik, *J. Electrochem. Soc.* **129** (1982) 1973-1977

[17]　K. Azumi, T. Ohtsuka, N. Sato, *J. Electrochem. Soc.* **134** (1987) 1352-1357

[18]　U. Stimming, *Electrochim. Acta* 31 (1986) 415-429

[19]　L. Hamadou, A. Kadri, N. Benbrahim, *Corros. Sci.* **52** (2010) 859-864

[20]　F. Di Quarto, M. Santamaria, *Corros. Eng. Sci. Techn.* **39** (2004) 71-81

[21]　Y.F. Cheng, J.L. Luo, *Electrochim. Acta* **44** (1999) 2947-2957

4

Influence of ceramic separator's characteristics on microbial fuel cell performance

Anil N. Ghadge, Mypati Sreemannarayana, Narcis Duteanu* and
Makarand M. Ghangrekar[✉]

Department of Civil Engineering, Indian Institute of Technology, Kharagpur -721302, India
*University "Politehnica" of Timisoara, Industrial Chemistry and Environmental Engineering,
2 Victoria Sq., 300006 Timisoara, Romania

[✉]Corresponding Author

Abstract

This study aimed at evaluating the influence of clay properties on the performance of microbial fuel cell made using ceramic separators. Performance of two clayware microbial fuel cells (CMFCs) made from red soil (CMFC-1) typically rich in aluminum and silica and black soil (CMFC-2) with calcium, iron and magnesium predominant was evaluated. These MFCs were operated under batch mode using synthetic wastewater. Maximum sustainable volumetric power density of 1.49 W m^{-3} and 1.12 W m^{-3} was generated in CMFC-1 and CMFC-2, respectively. During polarization, the maximum power densities normalized to anode surface area of 51.65 mW m^{-2} and 31.20 mW m^{-2} were obtained for CMFC-1 and CMFC-2, respectively. Exchange current densities at cathodes of CMFC-1 and CMFC-2 are 3.38 and 2.05 times more than that of respective anodes, clearly indicating that the cathodes supported much faster reaction than the anode. Results of laboratory analysis support the presence of more number of exchangeable cations in red soil, representing higher proton exchange capacity of CMFC-1 than CMFC-2. Higher power generation was observed for CMFC-1 with separator made of red soil. Hence, separators made of red soil were more suitable for fabrication of MFC to generate higher power.

Keywords

Cation exchange capacity; Coulombic efficiency; Charge transfer resistance; Charge transfer coefficient; Exchange current density; Power density; Wastewater treatment

Introduction

Recently considerable attention is being paid on the two major problems of the world, which are namely maintaining quality of water body and energy crisis. Solution to these problems could be provided by microbial fuel cell (MFC) to treat organic matter present in wastewater and simultaneously produce bio-electricity [1-5]. Although, considerable progress has been achieved in the performance of a MFC in the past ten years, one of the main challenge for commercializing scalable MFCs is the high cost and low mechanical strength of the separator materials used for fabrication of this device. Tian *et al.* [6], demonstrated that placing of an anaerobic membrane filtration process sequentially with an MFC accomplished efficient nutrients removal with low propensity of membrane fouling. It was reported that the use of poly(tetrafluoro-ethylene) (PTFE) layered activated charcoal electrode and Zirfon® as separator, improved MFCs performance and can be used to replace costly polymeric membrane and expensive catalyst in MFCs [7]. Proton exchange membrane (PEM) such as Nafion [8], nano-composite membrane made of sulfonated polymer (ether ether ketone) and Montmorillonite Clay [9], nano-composite membranes of Nafion and montmorillonite clay [10] were used in the MFCs to separate anodic chamber from cathodic chamber. However, these polymeric membranes or composite membranes are costly; hence, limit the practical application of the MFCs.

Ceramic membranes are found to be promising materials in MFCs because of their low production cost and better structural strength, thus, providing an alternative for the costly polymeric membranes [11-15]. Application of such ceramic membranes in MFC has been practiced since last ten years and its utility has been demonstrated through different studies. This was the first attempt where Park and Zeikus [11] developed a porcelain septum separator for single chambered MFC using 100% Kaolin and found comparable performance with that dual chambered MFC. A three layered cathode composed of a cellulose acetate film, a ceramic membrane, and a porous graphite plate to create a single chamber MFC that linked with solar cell to enhance power generation [12]. Behera and Ghangrekar [13] studied the effect of different thickness of such ceramic membrane on performance of the dual chambered MFC, and reported better power output for MFC having smallest thickness of the membrane. Use of terracotta pot for making single chamber MFC, after coating outer surface with conductive graphite paint, demonstrated Coulombic efficiency of 21 ± 5 % with power density of 33.13 mW m^{-2} [14]. More recently, Winfield *et al.* [15] compared performance of MFCs made from terracotta and earthenware by considering wall thickness, porosity and cathode hydration. More porosity of earthenware proved to be the better material compared to terracotta. However, these studies do not include the effect of soil pH, conductivity and cation exchange capacity (CEC) on the performance of MFCs.

For effectual use of such ceramic membranes, made from clay minerals, they should have higher cation exchange capacity. The existence of the pH dependent charge portion of the cation exchange capacity of soils is widely accepted for many years [16]. Electrical conductivity of the soil is the measure of salt concentration in the soil solution. Bulk electrical conductivity of soil is generally assumed to be dominated by the electrical conductivity of the soil solution, with perhaps a small contribution from surface charges associated with soil solids [17].

In MFCs, the rate of proton consumption at the cathode is often higher than the transfer rate through the membrane [18,19]. Hence, for enhancing power generation of this device the separator used should offer higher rate of proton/cation transfer. The transfer of protons from a protonated species to an uncharged molecule at the surface of the clay mineral is an important process [20]. The soil used for making ceramic separator in MFC participates in exchange of cations from anodic chamber to cathodic chamber. Protons released during the oxidation of organic matter from the anodic chamber are being adsorbed on to the surface of the soil by replacing the loosely held cations. The layered silicate clay minerals like smectite clays, show attractive hydrophilic properties and good thermal stability at high temperature [21]. The layered silicates commonly used for proton exchange membrane fuel cell applications are montmorillonite made of silica tetrahedral and alumina octahedral sheets which has advantageous hygroscopic properties [22,23]. The cations Ca^{2+}, Mg^{2+}, K^+ and Na^+ are called the base cations and H^+ and Al^{3+} are called acidic cations. The acidity of the soil is the amount of the total cation exchange capacity (CEC) occupied by the acidic cations [24]. More than proton, this cation migration also affects the performance of cathode, hence overall performance of MFC.

Porosity of the soil represents the hydraulic conductivity which depends upon the pore throat radii of clay materials. Typically clays have very low hydraulic conductivity due to their small throat radii. For MFC made with such clayware separator, different soil porosities play a vital role in the seepage of substrate from anodic to cathodic chamber [25]. Apart from loss of fuel, this may lead to the availability of organic matter at higher concentration on the cathode, supporting heterotrophic bacterial growth on cathode and thereby reducing cathode potential. Under such circumstances, the cathode often gives negative potentials (vs. Ag/AgCl), than the positive potential it is expected to give, while using oxygen as an electron acceptor [26]. Hence, hydraulic conductivity and cation exchange capacity are the important properties of the materials to be selected as a separator in MFC.

The objective of this study was to investigate the effect of different soil properties like pH, conductivity, porosity, cation exchange capacity on the performance of MFCs having ceramic separators made from two different soils. In addition, the electrode reaction kinetics was investigated for assessing performance of these MFCs.

Experimental

Construction of Microbial Fuel Cell

The study was carried out using dual-chambered Clayware Microbial Fuel Cells (CMFCs). The anodic chambers of these CMFCs were made up of baked clayware pot and the wall material of the pot (about 5 mm thick) itself acted as a separator allowing transfer of protons from anode to cathode. The pots were made from the red soil (typically rich in aluminum and silica) in CMFC-1 and black soil (rich in calcium, iron and magnesium predominant) in CMFC-2. The anodic chamber of the CMFC-1 and CMFC-2 had a liquid volume capacity of 550 ml and 700 ml, respectively. Cathodic chambers in both the MFCs were made up of plastic container having 5 liter capacity. Although there is difference in anodic chamber volume of both the MFCs, however, cathodic chamber volume of 5 litre, which was kept same in both the MFCs. It is important to note here that the rate of proton transfer largely depends on the separator area to anodic chamber volume ratio (S_a/v). In the present study, this ratio was 83.3 and 86.5 m^2/m^3, respectively, for CMFC-1 (separator made of red soil) and CMFC-2 (separator made of black soil), which indicates that there was no significant difference in S_a/v ratio. Carbon Felt (Panex®35, Zoltek Corporation) with 230 cm^2 and 261 cm^2 projected surface areas were used as cathode in CMFC-1 and CMFC-2, respectively. Anodes in CMFC-1 and CMFC-2 were made from stainless steel mesh having total surface area of 268 cm^2 and 304 cm^2, respectively. An aquarium aerator was inserted at the bottom of cathodic chamber to supply air continuously with an aquarium air pump (SOBO Aquarium Pump, China). The connections between two electrodes were made with concealed copper wire through external resistance of 100 Ω.

Inoculation and operation of CMFCs

Anaerobic mixed sludge collected from septic tank was used as an inoculum in the anodic chamber of the CMFCs. When mixed anaerobic sludge is used as source of inoculum, it contains both electrogenic as well as non-electrogenic (mostly methanogenic) bacteria. In the anodic chamber of MFC, it is necessary to dominate electrogenesis to obtain higher Coulombic efficiency. Methanogens in the MFCs compete for substrate and electrode space with electrogenic bacteria and reduce the power output. Therefore, the inoculum sludge was given a heat pre-treatment (heated at 100 °C for 15 min) to suppress methanogens and required amount of sludge was added to the anodic chamber [27]. Synthetic wastewater containing sodium acetate as a source of carbon with chemical oxygen demand (COD) of about 3000 mg L^{-1} was used in this study. The sodium acetate medium was prepared by adding 3843 mg L^{-1} CH_3COONa, 4500 mg L^{-1} $NaHCO_3$, 954 mg L^{-1} NH_4Cl, 81 mg L^{-1} K_2HPO_4, 27 mg L^{-1} KH_2PO_4, 750 mg L^{-1} $CaCl_2.2H_2O$, 192 mg L^{-1} $MgSO_4.7H_2O$ and trace metals like Fe, Ni, Mn, Zn, Co, Cu, and Mo as per the composition given by [27]. The feeding frequency of 5 days was adopted. These CMFCs were operated at temperatures varying from 33 to 37 °C under batch mode. The pH of tap water used as catholyte remained in the range of 8.2-8.5; whereas, anolyte pH was in the range of 7.1-7.4.

Analysis and calculations

The pH and conductivity of anolyte and catholyte was measured using pH meter (Cyber Scan pH 620) and TDS meter (Cyber Scan CD 650, Eutech instruments, Singapore), respectively. COD concentrations were measured according to APHA standard methods [28], using closed reflux method. The performance of CMFCs was evaluated in terms of voltage (*U*) and current (*I*) measured using a digital multimeter with data acquisition unit (Agilent Technologies, Malaysia) and converted to power according to *P* = *UI*, where *P* = power, W; *I* = current, A; and *U* = voltage, V. Power density and power per unit volume were calculated by normalizing power to the anode surface area and net liquid volume of anodic chamber, respectively. The current density i_d was calculated using

$$i_d = \frac{U}{R_{ext} A_d}$$

(1)

where, R_{ext} is the external resistance (Ω) and A_d (m^2) is the surface area of the anode. Polarization studies were carried out by varying the external resistance from 10000 to 10 Ω using the resistance box (GEC 05 R Decade Resistance Box) and cell voltages (*U*) were recorded. Internal resistance of the CMFCs was measured from the slope of the line from plot of voltage versus current [29]. Columbic Efficiency (*CE*) was determined by integrating the current measured over time, *t*, and compared with the theoretical current on the basis of COD removal and calculated as [1]:

$$CE = \frac{M \int_0^t I dt}{F b V \Delta COD}$$

(2)

where, *V* is the volume of the anodic chamber of MFC; *M* = 32, molecular weight of oxygen; *F*, Faraday's constant = 96485 C mol^{-1}; *b* = 4, the number of electrons exchanged per mole of oxygen; *ΔCOD* is the difference in the influent and effluent COD for time *t*.

Analysis of the soil properties

The pH and the conductivity of the soil samples were measured according to the Indian Standard method of test for soils. Indian standard IS: 2720 (Part 26) – 1987 was used for determination of pH value [30] and conductivity of the soil was measured according to IS 14767: 2000 [31]. Cation exchange capacity of the soil was measured according to Indian standard, IS: 2720 (Part 24) – 1976 [32]. Chemical constituents for the red and black soils were obtained through the X-Ray Fluorescence (XRF) analysis. Porosity of the soil was indirectly measured from the percentage water absorbed by the clayware pot made from respective soils after immersing in water for 24 hours.

Reaction kinetics at electrodes

Tafel plot, as derived from equation (3) [33], was employed to measure the reaction kinetics for working electrode (anode and cathode) and Ag/AgCl was used as the reference electrode. The reference electrode was placed in the working chamber during the measurements.

$$\ln\left(\frac{i}{i_0}\right) = \beta\left(\frac{F\eta}{RT}\right)$$

(3)

where i_0 is exchange current density, i is the electrode current density (mA m^{-2}), α is the electron transfer coefficient, R is the ideal gas constant (8.31 J mol^{-1} K^{-1}), F is the Faraday's constant (96,485 C mol^{-1}), T is the absolute temperature, K and η is the activation overpotential. Purpose of using Tafel plot is to calculate the i_0 and α value. The i_0 is a fundamental parameter in the rate of electro-oxidation or electro-reduction of a chemical species at an electrode at equilibrium. The charge transfer resistance (R_{ct}) was calculated from the following equation:

$$R_{ct} = \frac{RT}{nFi_0}$$

(4)

where, n is the number of electrons.

Results and Discussion

Physico-chemical properties of the soil used in CMFCs

The soils used for manufacturing the pots showed different pH. The electrical conductivity and cation exchange capacity of the red soil is higher than that of the black soil, indicating usefulness of the former in the clayware separator application (Table 1). However, the porosity of the pot made from black soil was higher than the pot made from red soil. Higher porosity may allow the anolyte to come to the cathode resulting in not only the physical substrate loss but also it will allow oxygen to penetrate in anodic chamber, reducing Coulombic efficiency of the system due to direct oxidation of the substrate.

Table 1. Physical, chemical and electrical properties of red and black soil used for making separators

Sl. No	Soil properties	Red soil (CMFC-1)	Black soil (CMFC-2)
1	pH	7.4	8.5
2	Porosity, %	11.6	17.6
3	Electrical conductivity, mS cm^{-1}	2.403	0.045
4	Cation exchange capacity (CEC), mmol (kg soil)$^{-1}$	125	20

Wastewater treatment

After inoculating the anodic chamber of the CMFCs with heat pretreated anaerobic mixed consortia, synthetic feed was supplied and wastewater treatment performance of the CMFCs under different feed cycles was observed. Average COD removal efficiency of 78.9 ± 3.9 % and 89.6 ± 3.2 % was observed in CMFC-1 and CMFC-2, respectively. COD removal efficiency in the CMFC-2 was higher than the CMFC-1. It was observed that porosity of clayware pot used in CMFC-2 was 52 % higher (Table 1) than that of CMFC-1, because of which probably it has permitted more diffusion of oxygen from cathodic chamber to anodic chamber to support aerobic oxidation of fraction of substrate present in anodic chamber to establish higher COD removal efficiency. The oxygen diffusion coefficient in the range of 5.38×10^{-6} to 6.67×10^{-6} cm^2 s^{-1} is reported in early studies for this clayware separator by Behera and Ghangrekar [13]. In addition, due to more porosity of separator used in CMFC-2, exchange of water molecules due to osmosis across the membrane might have diluted the anolyte, resulting in higher COD removal rate.

Electricity generation

Performance of CMFCs was evaluated by measuring the open circuit voltage and operating voltage. The current and voltage gradually increased with time of operation. The maximum voltage across 100 Ω resistance of 286 mV and 280 mV was observed in CMFC-1 and CMFC-2, respectively. CMFC-1 generated a maximum sustainable power density (normalized to the anode surface area) and volumetric power (normalized to the working volume of anodic chamber) of 30.5 mW m^{-2} and 1.49 W m^{-3} (Fig. 1), respectively; whereas, CMFC-2 generated power density of 25.7 mW m^{-2} and volumetric power of 1.12 W m^{-3}. The power produced by CMFC-1, made from red soil, was 1.33 times higher than the CMFC-2 wherein the separator was made from black soil with lower CEC and electrical conductivity. It is interesting to note here that in spite of having higher separator area and more liquid volume (anodic chamber) for CMFC-2, it generated less power compared to CMFC-1.

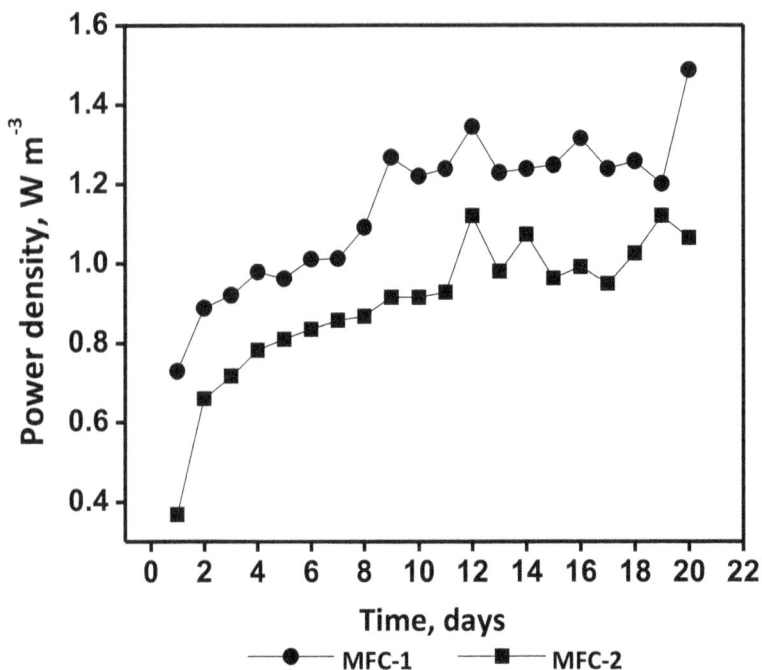

Figure 1. *Volumetric power density of CMFC-1 and CMFC-2*

Effect of chemical properties of soil used for making separator on electricity generation

The CEC of red soil is 6.25 times (Table 1) higher than black soil, indicating more number of exchange sites are available for the transfer of cations in red soil. Due to availability of more exchange sites in CMFC-1, better transfer of the protons occurred to improve the power generation in CMFC-1 compared to CMFC-2. In addition, the XRF data (Table 2) confirms that the aluminum content of the red soil is more than the black soil which makes the red soil more acidic than black soil. The pH of the red soil (Table 1) was lower than black soil confirming that the red soil is more acidic and has high capacity to hold the H^+ ions which improved the performance of CMFC-1 in terms of power generation.

Electrical conductivity is the measure of salt concentration in the soil solution. Soils high in smectite often exhibit high electrical conductivity due to water associated with the clays. The soil with high montmorillonite mineral can act as better proton exchange material due to its hydrophilic nature and the high cation exchange capacity [34]. The conductivity of soil used as separator in CMFC-1 is almost 53.4 times (Table 1) more compared to CMFC-2, authenticating utility of the red soil for making separator to harvest more power from the CMFCs.

Table 2. Chemical compounds present in red and black soil

Sl. No	Compound	Content, %		Sl. No	Compound	Content, %	
		Red soil	Black soil			Red soil	Black soil
1	Na_2O	3.95	0.273	14	Co	0.406	0.441
2	MgO	0.654	3.86	15	Ni	0.004	0.006
3	Al_2O_3	26.3	21.6	16	Cu	0.274	0.282
4	SiO_2	57.5	53.4	17	Zn	0.005	0.023
5	P_2O_5	1.13	0.204	18	Ga	0.001	0.001
6	SO_3	0.258	0.162	19	Rb	0.007	0.008
7	K_2O	1.78	0.798	20	Sr	0.004	0.017
8	CaO	0.791	10.4	21	Y	0.004	0.002
9	Fe_2O_3	4.70	6.75	22	Zr	0.013	0.010
10	Cl	1.45	0.071	23	Nb	0.001	0.0005
11	Ti	0.658	1.45	24	Ba	0.012	0.020
12	Cr	0.01	0.01	25	Ce	0.021	0.023
13	Mn	0.067	0.093	26	Pb	0.003	0.002

Coulombic efficiency

Coulombic efficiency compares the recovery of the coulombs through the external circuit against theoretical coulombs that is present in the organic matter. CMFC-1 showed average *CE* of 7.69 ± 1.52 %, whereas in CMFC-2 it was 6.39 ± 1.40 %. Higher *CE* of CMFC-1 than CMFC-2 might have been due to the difference in the CEC of red and black soil and also due to more diffusion of oxygen in case of black soil due to high porosity. In MFCs higher *CE* is reported with pure inoculum culture and with synthetic wastewater [35].

Polarization and Internal resistance

Polarization curve helps to understand the performance of MFC in terms of power generation and internal resistance. It represents the cell voltage and power density as a function of the current density. Figure 2 shows power and polarization curves obtained using variable resistor box for CMFC-1 and CMFC-2.

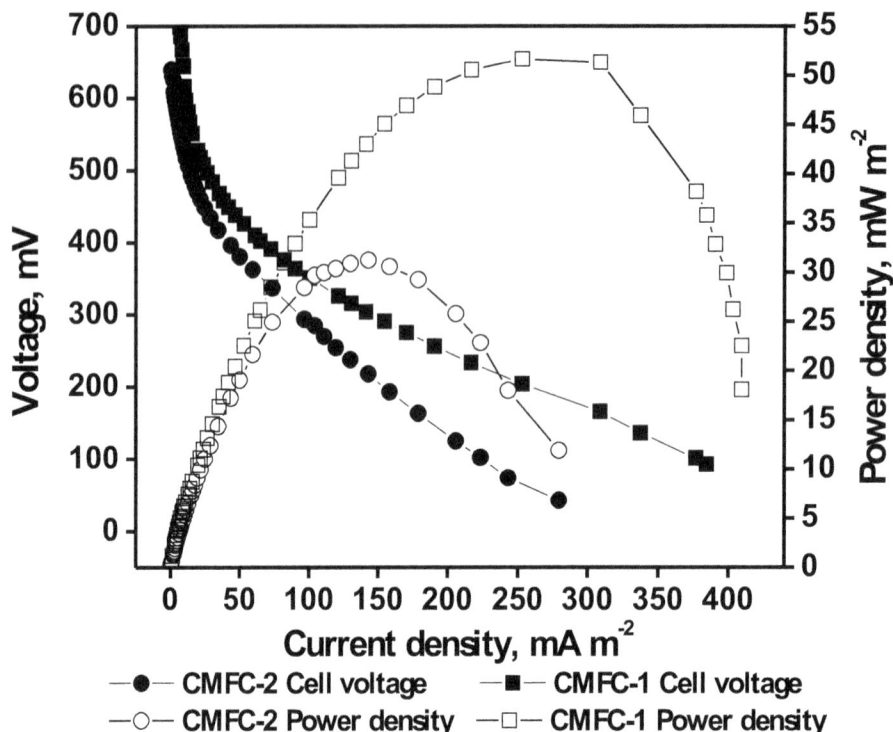

Figure 2. *Polarization curve for CMFC-1 and CMFC-2*

During polarization, the maximum power density observed for CMFC-1 was 51.65 mW m^{-2} (E = 0.204 V, R_{ext} = 30 Ω) and that of CMFC-2 it was 31.20 mW m^{-2} (E = 0.217 V, R_{ext} = 50 Ω). This indicates that the higher CEC of red soil supported better proton transfer from the anode to the cathode in CMFC-1. Conversely, lower power output observed in CMFC-2 could be attributed to the lesser CEC of black soil used for making separator (Table 1). It is well documented that the cation exchange capacity of soil plays vital role in the proton transfer mechanism in soil [36].

Internal resistances of CMFC-1 and CMFC-2 measured from the slope of the plot of voltage vs. current were 36 Ω and 56.5 Ω, respectively. In the region of low current density (Fig. 2) rapid voltage drops were observed in both the MFCs, and in the region of high current density, voltage decreased linearly at lower rate. Lower proton transfer rate and low conductivity of black soil used for making separator of CMFC-2 increased the internal resistance.

Electrode Potential

Electrode potentials represent the energy level of the electrons at anode and cathode. Electrons move from area of higher potential energy to area of lower potential energy. As the anode has a higher potential energy so electrons move from anode to cathode through an external circuit. During polarization, cathode of CMFC-1 well supported for the reduction reaction up to 0.5 mA current at 1000 Ω external resistance. However, the cathode potential of CMFC-2 dropped to zero (vs. Ag/AgCl) at 0.25 mA current at 2100 Ω external resistance, showing inefficiency of cathode for reduction reaction at higher current (Fig. 3).

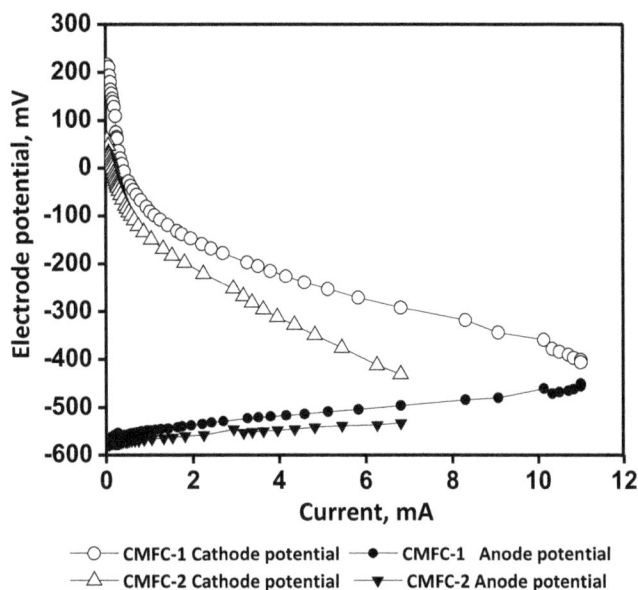

Figure 3. Change of the cathode and anode potentials during polarization in CMFC-1 and CMFC-2

The open circuit potentials (OCP) for anode observed before polarization (*vs.* Ag/AgCl) for CMFC-1 and CMFC-2 were -610 mV and -580 mV, respectively. During polarization, increase in anode potentials was observed in both the MFCs due to transfer of electrons from anode to cathode, thus positive overpotential was observed. However, the anode potentials during polariz-ation in both the MFCs were only slightly increased, indicating better stability of the anodes.

Electrode Kinetics

Tafel plot analyses were carried out to determine the exchange current density (i_0), charge transfer coefficient (α) and charge transfer resistance (R_{ct}). The values obtained from these analyses are summarized in Table 3. Based on the Tafel-type linear equation obtained from the graphs (Figs. 4A, 4B), the slope is $\alpha F/RT$ and the y-axis intercept is the logarithm of the exchange current.

For anodic reaction at 25°C, the slope of Tafel plot is

b = 0.059 / 1 - α (5)

and at the same time the slope for cathodic reaction is

b = 0.059 / α(6)

The value of i_0 represents the rate of exchange current density at equilibrium state when the reaction overpotential is zero. Higher the exchange current (i_0) faster is the reaction rate, resulting in a lower activation energy barrier of forward reaction [37]. The electrode materials used in both the CMFCs were same, however different reaction kinetics at electrodes was observed. The reactions at cathodes were faster than anode. Presence of very high actual surface area of the carbon felt material resulted in producing a low cathodic overpotential [38]. Comparing the i_0 values for the reduction reactions at cathode of CMFC-1 and CMFC-2, the CMFC-1 with separator made from red soil had better performance. It indicates that the reactions at cathode of CMFC-1 were faster; might be due to the higher transfer of H^+ ions and other cations in CMFC-1, enhancing the reaction rates at cathode. Comparing the anodes of both the CMFCs, reactions at the anode of CMFC-2 were slightly faster than CMFC-1. Apart from the differences in the CEC, the reactions at cathode of CMFC-2 were slower than CMFC-1. This could be probably due to the higher porosity of the clayware separator used in CMFC-2, due to which the substrate exchange occurred and oxygen supplied in the cathodic chamber was utilized by the substrate. The exchange current densities of CMFC-1 and CMFC-2 cathodes were 3.38 and 2.05 times more than that of the respective anodes, clearly indicating that the reaction at cathode was much faster than anode.

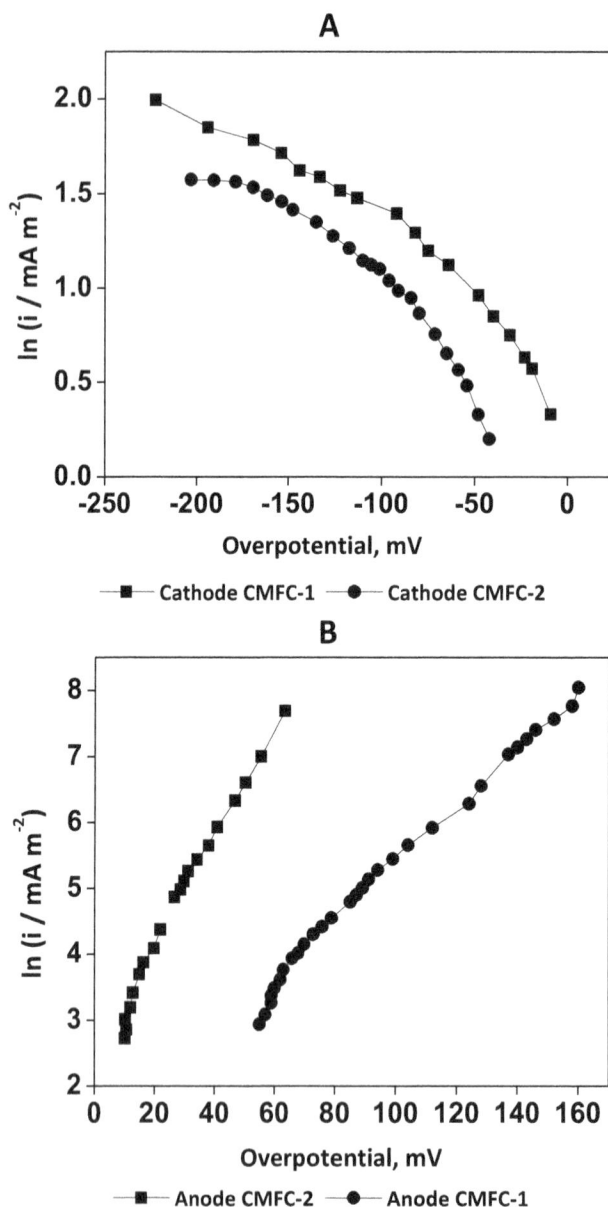

Figure 4. Tafel plots for **A** - cathode of CMFC-1 and CMFC-2, and
B - anode of CMFC-1 and CMFC-2

According to the Butler–Volmer model of electrode kinetics, the charge transfer coefficient (α) is used to describe the symmetry between the forward and reverse electron transfer steps and the magnitude of α ranges in value from 0 to 1. Charge transfer coefficient signifies the fraction of the interfacial potential at an electrode-electrolyte interface that helps in lowering the free energy barrier for the electrochemical reaction. Lower electron transfer coefficient indicates less activation energy required for the electron transfer, resulting lower activation loss [37]. Charge transfer resistance (R_{ct}) represents the capability to resist the transfer of charge from electrode-electrolyte interface. It is interesting to note that the R_{ct} and α (Table 3) of cathodes for both the MFCs were much lesser than the anode of both the MFCs.

Table 3. *Tafel analysis of CMFC-1 and CMFC-2*

Parameter	CMFC-1		CMFC-2	
	Anode	Cathode	Anode	Cathode
Exchange current density (i_0), mA m^{-2}	0.60	2.03	0.74	1.52
Charge transfer coefficient (α)	0.30	0.032	0.45	0.038
Charge transfer resistance (R_{ct}), Ω m^2	10.70	3.16	8.67	4.22

The R_{ct} and α values for cathode of CMFC-1 were lower than CMFC-2, which supports that the clayware membrane made from red soil supported better reaction at the cathode. This is because of high cation transported from CMFC-1 to the cathode side, increased the rate of electrochemical transformation with lower electrical energy loss, thus charge transfer coefficient gets reduced. Lower R_{ct} for anode of CMFC-2 than anode of CMFC-1 indicated that the anode of CMFC-2 was performing slightly better, as also evident from the exchange current density. However, due to limitations of the cathodic reactions the overall performance of CMFC-2 was inferior as compared to CMFC-1.

Conclusions

Properties of the clayware separator such as CEC, pH and electrical conductivity influenced the performance of MFCs. The power generation of MFC having separator made from red soil was better than the black soil, due to high CEC, low pH and higher electrical conductivity of the red soil. Results of Tafel plots showed that lower exchange current density and higher charge transfer resistance of anodes compared to cathodes, contributed towards more activation loss in both the MFCs. In spite of similar electrode materials in both CMFCs, variation in electrode kinetics accentuate effect of properties of separator on the performance of CMFCs. Detailed studies on the mineral composition of soils are required to enhance the CEC for further improving power generation of MFC made with such low cost clayware separator. Development of such efficient and cheaper separator material will help in drastically reducing fabrication cost of MFC for field implementation.

Acknowledgement: Grants received from Department of Science and Technology, Government of India (File No. DST/TSG/NTS/2010/61) to undertake this work is duly acknowledged.

References

[1] B. E. Logan, B. Hamelers, R. Rozendal, U. Schröder, J. Keller, S. Freguia, P. Aelterman, W. Verstraete, K. Rabaey, *Environ. Sci. Technol.* **40** (2006) 5181-5192.

[2] D. A.Lowy, L. M. Tender, J. G. Zeikus, D. H. Park, D. R. Lovley, *Biosens. Bioelectron.* **21** (2006) 2058-2063.

[3] K. Rabaey, W. Ossieur , M. Verhaege, W. Verstraete, *Water Sci. Technol.* **52(1)** (2005) 515-523.

[4] K. Rabaey, W. Verstraete, *Trends Biotechnol.* **23** (2005) 291-298.

[5] B. E. Logan, J.M. Regan, *Environ. Sci. Technol.* **40** (2006) 5172-5180.

[6] Y. Tian, C. Ji, K. Wang, P. Le-Clech, *J. Membr. Sci.* **450** (2014) 242-248.

[7] D. Pant, G. Van Bogaert, M. De Smet, L. Diels, K. Vanbroekhoven, *Electrochim. Acta* **55** (2010) 7710-7716.

[8] R. A. Rozendal, H. V. Hamelers, C. J. Buisman, *Environ. Sci. Technol.* **40(17)** 5206-5211.

[9]   M. M. Hasani-Sadrabadi, S. H. Emami, R. Ghaffarian, H. Moaddel, *Energ. Fuel* **22** (2008) 2539-2542.

[10]   C. Felice, S. Ye, D. Qu, *Ind. Eng. Chem. Res.* **49** (2010) 1514-1519.

[11]   D. H. Park, J. G. Zeikus, *Biotechnol. Bioeng.* **81** (2003) 348-355.

[12]   H. N. Seo, W. J. Lee, T. S. Hwang, D. H. Park, *J. Microbiol. Biotechnol.* **19** (2009) 1019-1027.

[13]   M. Behera, M. M. Ghangrekar, *Water Sci. Technol.* **64** (2011) 2468-2473.

[14]   F. F. Ajayi, P. R. Weigele, *Bioresour. Technol.* **116** (2012) 86-91.

[15]   J. Winfield, J. Greenman, D. Huson, I. Ieropoulos, *Bioprocess Biosyst. Eng.* **36** (2013) 1913-1921.

[16]   V. V. Volk, M. L. Jackson, *Clays Clay Mineral.* **12** (1964) 281-285.

[17]   A. G. Hunt, S. D. Logsdon, D. A. Laird, *Soil Sci. Soc. Am. J.* **70** (2006) 14-23.

[18]   G. C. Gil, I. S. Chang, B. H. Kim, M. Kim, J. K. Jang, H. S. Park, H. J. Kim, *Biosens. Bioelectron.* **18** (2003) 327-334.

[19]   H. Liu, B. E. Logan, *Environ. Sci. Technol.* **38** (2004) 4040-4046.

[20]   K. Raman, M. Mortland, *Soil Sci. Soc. Am. J.* **33** (1969) 313-317.

[21]   M. F. Delbem, T. S. Valera, F. R. Valenzuela-Diaz, N. R. Demarquette, *Quím. Nova* **33** (2010) 309-315.

[22]   J. H. Chang, J. H. Park, G. G. Park, C. S. Kim, O. O. Park, *J. Power Sources* 124 (2003) 18-25.

[23]   B. Liao, M. Song, H. Liang, Y. Pang, *Polymer* **42** (2001) 10007-10011.

[24]   J. Derome, A. J. Lindroos, *Environ. Pollut.* **99** (1998) 225-232.

[25]   G. P. Matthews, G. M. Laudone, A. S. Gregory, N. R. A. Bird, A. D. G. Matthews, W. R. Whalley, *Water Resour. Res.* **46** (2010) W05501.

[26]   M. Behera, P. S. Jana, M. M. Ghangrekar, *Bioresour. Technol.* **101** (2010) 1183-1189.

[27]   G. S. Jadhav, M. M. Ghangrekar, *Appl. Biochem. Biotech.* **151** (2008) 319-332.

[28]   APHA, *Standard methods for examination of water and wastewater.* American Public Health Association, American Water Works Association, Water Environment Federation, 20th ed.,Washington, DC 1998.

[29]   C. Picioreanu, I. M. Head, K. P. Katuri, M. van Loosdrecht, K. Scott, *Water Res.* **41** (2007) 2921-2940.

[30]   Indian standards, IS:2720-Part 26, *Methods of test for soils (Determination of pH value)*, Bureau of Indian Standard, New Delhi, India 1987.

[31]   Indian standards, IS:14767, *Methods of test for soils (Determination of specific electrical conductivity)*, Bureau of Indian Standard, New Delhi, India, 2000.

[32]   Indian standards, IS:2720-Part 24, *Methods of test for soils (Determination of cation exchange capacity)*, Bureau of Indian Standard, New Delhi, India, 1976.

[33]   A. J. Bard, L. R. Faulkner, *Electrochemical methods: principles and applications*, John Wiley & Sons, Inc. New York, 2001.

[34]   H. Quiquampoix, *J. Soil Sci. Plant Nutr.* **8** (2008) 75-83.

[35]   P. Aelterman, K. Rabaey, H. T. Pham, N. Boon, W. Verstraete, *Environ. Sci. Technol.* **40** (2006) 3388-3394.

[36]   B. Ulrich, *J. Plant Nutr. Soil SC.* **149** (1986) 702-717.

[37]   S. Srikanth, M. Venkateswar Reddy, S. Venkata Mohan, *Bioresour. Technol.* **119** (2012) 241-251.

[38]   Q. Deng, X. Li, J. Zuo, A. Ling, B.E. Logan, *J. Power Sources* **195** (2010) 1130-1135.

Preparation and properties of electrodeposited Ni-TiO$_2$ composite coating

Sukhdev Singh Bhogal$^{\boxtimes}$, Vijay Kumar*, Sukhdeep Singh Dhami and Bahadur Singh Pabla

National Institute of Technical Teachers and Research, Sector-26, Chandigarh, India
*Punjab Technical University Campus, SAS Nagar (Mohali)-Punjab, India

$^{\boxtimes}$ Corresponding Author

Abstract

The mechanical properties of cutting tools, such as microhardness, corrosion resistance, and coating adhesiveness, directly affect the tool life and indirectly affect the component cost. In this paper, Ni-TiO$_2$ composite coating was prepared through electrocodeposition in order to improve the mechanical properties of tungsten carbide cutting tools. The microhardness of the Ni-TiO$_2$ composite layer was studied by varying the input current density (mA cm^{-2}), pH value of the electrolyte, and particle concentration of TiO$_2$ in electrolyte bath. The microstructure and phase structure of the composite layer were investigated using atomic force microscopy , scanning electronic microscopy and X-ray diffraction. The surface morphology of the Ni-TiO$_2$ coated layer shows fine-grained structures and higher microhardness at lower currents. The maximum microhardness of the coated layer, 1483 HV, is found at a current of 15 mA cm^{-2} and Watts solution pH of 4.5. It is observed that with the increase of TiO$_2$ content, the microhardness of the coating also increases.

Keywords

Microhardness; Electrocodeposition; Ccomposite layers

Introduction

TiO$_2$ is well known as a metal dioxide ceramic which has been utilized in many applications for its fine physical and chemical properties. Nano-size TiO$_2$ particles are usually introduced into nickel-based composite coating to enhance its properties such as corrosion resistance, microhardness, and abrasiveness [1]. Electrodeposition is one of the traditional processes used for the improvement of the surface. Specially, it is used for the enhancement of coating properties such decorativeness, functionality, and electroforming of the coated surfaces. This method has advantages like low cost, low temperature, ease of use, and being a single-step process that does not require additional thermal treatment [2]. In the manufacturing industry, metal cutting tools are in high demand due to their greater sustainably in the cutting of hard materials, as tough conditions of high cutting force and high contact pressure between the workpiece and tool materials are found during machining. Further cutting conditions and tool shape, make the tool condition to the severity and ultimately causes the earlier tool failure. Mostly the cutting tools which are used have more ultimate resistance against the load and ability to limit the thermal and mechanical stresses during cutting of hard and difficult-to-cut materials. Nowadays, high-performance cutting tool materials obtained with chemical vapour deposition (CVD) and physical vapour deposition (PVD) as well as other coated carbides like cermets, ceramics, cubic boron nitride, and diamond-coated tools are frequently used, but tungsten carbide (WC)-coated tools are also in use for metal-cutting applications due to their high strength and the fact that they can be manufactured economically with complicated geometries.

Protective coatings on cutting tools have been in use for a little more than three decades. The search for wear-resistant materials has since been redirected to encompass material properties other than just hardness. Cutting performance is also dependent on both the cutting tool and the machine tool system, that is, on the tool material, cutting tool geometry, and hard coating [3]. In the early twenty century hard coatings were deposited by CVD and PVD by a magnetron sputtering process. Magnetron sputtering is preferred to the CVD process due to the contamination that occurs in the latter [4]. Single-layer coatings such as TiN, TiC, CrN, DLC, Ti-O_2, and AlN played an important role a few years ago, due to their mechanical and tribological properties [5]. Nowadays, these coatings are being mixed, designed, and improved by creating hetero-structured thin films and multilayered coatings with periods in the range of nanometres, which is one of the advanced coating concepts more commonly investigated during the last two decades, together with nanocomposites [6,7]. The nanocomposite coatings range from 3 to 10 nm in size, have been selected from nitrides, carbides, borides, and oxides, and are embedded in an amorphous or crystalline matrix. The large volume of grain boundaries provides ductility through sliding of the grain boundary along grain/matrix interfaces. These coatings have sometimes been reported to have hardnesses of up to 100 GPa [8,9].

Electrodeposition offers some unique advantages over other techniques such as improved appearance, corrosion resistance, and physiochemical properties of surfaces which find potential applications in the fabrication of advanced components for micro-/nano-electromechanical systems (MEMS/NEMS). Moreover the initial setup cost for coating by electrodeposition is quite low compared to other conventional coating processes like CVD and PVD. Moreover fine structure growth and easy control of the coating thickness up to fractions of a micrometre are possible via electrodeposition. The aim of this work is to evaluate the influence of Ni-TiO_2 composite coatings deposited onto a WC cutting tool by the electrodeposition technique. Ni-TiO_2 composite coating was employed to investigate the microhardness behaviour, adhesion to the substrate, and coating thickness with different currents at constant temperature and deposition time.

Experimental details

Ni-TiO_2 composites were electrodeposited from an organic free Watts nickel electrolyte with suspended TiO_2 nanoparticles on the WC tool bits by electrodeposition. Electrodeposited nickel and composite Ni-TiO$_2$coatings were prepared from Watts solution. The composition of the plating solution, as well as the plating conditions, is given in Table 1. All the chemicals used were of analytical grade with high purity and were purchased from Merck. A Gamry Reference 3000 potentiostat was used for coating. The triangular-shaped CNC tool bit (WC) was mechanically polished with different grades of emery papers (between 800 and 4000) and finally ultrasonically cleaned in acetone before electroplating without any activation pretreatment. The substrates were rinsed with doubly distilled water and dried under air.

A specially designed three-electrode electrodeposition cell was used for electrodeposition of Ni-TiO$_2$, where the cutting tool was the working electrode, Ag/AgCl was the reference electrode, and Pt wire acted as the counter electrode (anode). The distance between the Ag/AgCl reference electrode and the working electrode was 1 cm. The particle size of TiO$_2$ particles was about 50 nm. It was used without any pretreatment and was kept in a bath in suspension by continuous magnetic stirring at a rate of 210 rpm in order to maintain a uniform concentration of particles in the bulk solution. Before starting the electroplating, the electrolyte was placed in an ultrasonic bath for 20 min to prevent agglomeration of TiO$_2$ particles.

Table 1. *Watts solution for Ni/TiO$_2$ composite electrodeposits*

Electrolyte (Watts) solution composition	
Concentration of NiSO$_4$6H$_2$O, g L^{-1}	300
Concentration of NiCl$_2$6H$_2$O, g L^{-1}	45
Concentration of H$_3$BO$_3$, g L^{-1}	40
Concentration of TiO$_2$, g L^{-1}	6
pH of electrolyte	4.4 ± 0.1
Temperature, °C	30
Substrate	Tungsten carbide bit
Average current density, mA cm^{-2}	15, 30, 45
Agitation	180 rpm
Type of current	DC

The surface morphology of the Ni-TiO$_2$ composite coating was determined by SEM (JEOL, JSM-6460LV) with energy dispersive X-ray analysis (EDS) and AFM. To determine the phase structure of Ni–TiO$_2$ composite coatings, X-ray diffraction (XRD) analysis was performed on a PW 1710 diffractometer with Cu-Kα (λ = 1.54 A0) radiation. The operating target voltage was 40 kV and the tube current was 100 mA. Using the Scherrer equation, the average grain diameter could be calculated as follows:

$$D = \frac{K\lambda}{\omega\cos\theta} \tag{1}$$

where K is the constant of the grains (K = 0.89), λ the wavelength, ω the standard full width at half maximum (FWHM), and θ the Bragg angle. Vickers hardness was measured by applying a 0.5-kg force load for 10 seconds using a Vickers hardness tester (Mitutoyo).

Results and discussion

The surface morphology and microstructure of the composite coating were investigated through Atomic Force Microscopy (AFM) and Scanning Electron Microscopy (SEM). The surface morphology and microstructure of the coating layer are discussed below.

Surface morphology and microstructure

The composite coating was investigated by AFM. Figure 1 shows the AFM images of Ni-TiO$_2$ composite coating surfaces (20 × 20 µm area) on a tungsten carbide tool tip. Figures 1a and 1b show the surface after deposition of the Ni-TiO$_2$ coating with 5 mA of current and a pH of the solution of 4.5 respectively. In order to compare the surface roughness (*Ra*) of the coatings, AFM data are used, and based on the experimental results, the Ni-TiO$_2$ composite coating prepared at lower current i.e. at 15mA with pH 3.5 of solution (Fig. 1a) shows a uniform and fine structure among the micro-regions, whereas the coating prepared with a lower current and pH 4.5 of the Watts solution (Fig. 1b) appears to be finer and shows a regular crystalline grain structure. As the pH of the bath influences the hydrogen evolution voltage the precipitation of basic inclusion of solution and decomposition of the hydrated ion is deposited more freely. Thereby a finer and more uniform grain structure is obtained.

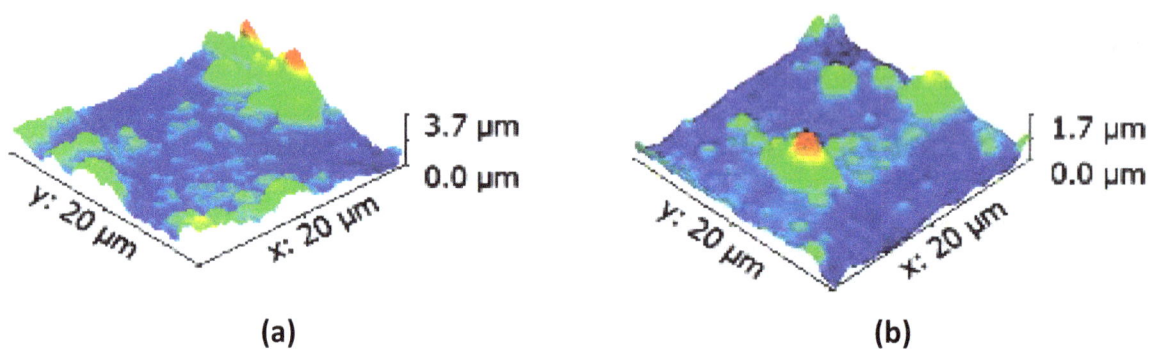

Figure 1. *AFM images of Ni-TiO$_2$ composite coating:* **(a)** *current = 15 mA, temperature = 30 °C, and pH 3.5;* **(b)** *current = 15 mA, temperature = 30 °C, and pH 4.5*

The XRD patterns of the Ni-TiO$_2$ (Fig. 2) composite coatings were detected by XRD to further confirm the existence of TiO$_2$ particles. Scans were recorded for the range 2θ = 20-80° with a scan step of 0.02°. The XRD patterns of the three types of Ni-TiO$_2$ composite coatings are presented in Fig. 2.

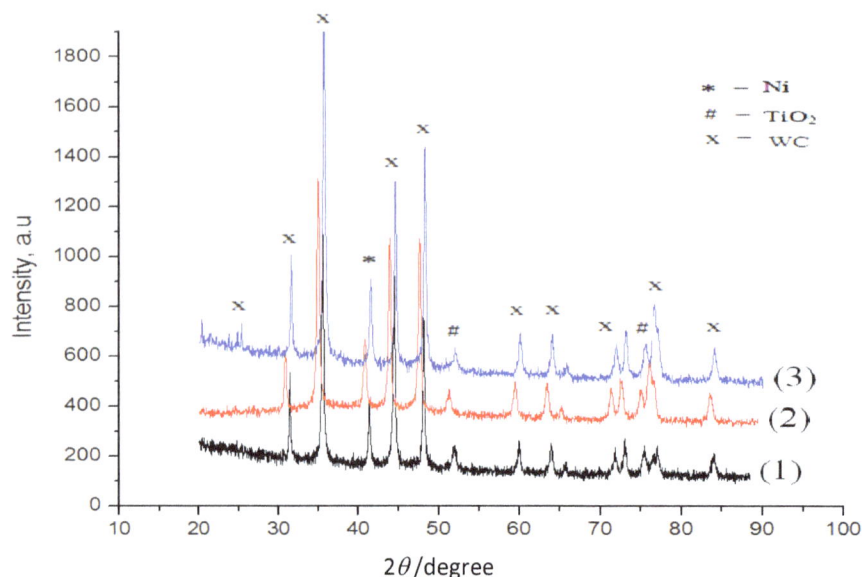

Figure 2. *XRD pattern of Ni-TiO$_2$ composite coating deposited at different conditions:* **(1)** *current = 15 mA and temperature = 30 °C;* **(2)** *current = 15 mA, temperature = 30 °C, pH 4.5* **(3)** *0.4 % TiO$_2$, current = 15 mA, temperature = 30 °C, pH 4.5*

The figure shows that the composite coatings consist of Ni phase and TiO_2 phase. For Ni, the diffraction peaks at 44.54°. For TiO_2, the diffraction peaks at 51.96° and 75.51°. According to the XRD data, the average grain size of Ni and TiO_2 can be calculated using Eq. (1). The results are shown in Table 2. The XRD results demonstrate the average grain diameters of Ni and TiO_2 in the composite coating prepared by electrocodeposition. At lower current, that is, 15 mA, the average grain sizes of Ni and TiO_2 are 60 and 19.65 nm, respectively, while at pH 4.5 the average grain sizes of the coating are 73.1 and 23.84 nm respectively. These results are consistent with the AFM results.

Table 2. *Average grain size of Ni and TiO_2 composite coatings*

Type of coatings	D_{Ni} / nm	D_{TiO_2} / nm
(1)	60.0	19.65
(2)	73.1	23.84
(3)	51.3	25.70

Mechanical properties of Ni- TiO_2 coatings

Mechanical properties such as microhardness play an important role in tool life enhancement. The inclusion of nano-sized TiO_2 particles in the metal matrix deposit is dependent on parameters like pH concentration, current density, temperature, stirring rate, and so on. The microhardness of the samples was examined by applying a 0.5-kg force load using a Vickers microhardness tester (Mitutoyo). After coating deposition, six indentations were made at different places and the mean of these values was taken. The effect of the applied current, the pH value of the Watts solution, and the percentage concentration of TiO_2 nanoparticles in solution by volume directly affected the microhardness of the composite coatings. The effects of these parameters are discussed below.

Effect of current on microhardness of Ni -TiO_2 coatings

Ni-TiO_2 composite coating is deposited on the cutting tool bit by varying the applied current from 15 to 45 mA. The effect of the current density applied during the process is studied by measuring the microhardness of the composite coatings. The morphological characterization of as-prepared samples was done using a scanning electron microscope (JEOL JSM 6510 LV) at an accelerating voltage of 20 kV as shown in Figure 3. The change in structure of the composite layer is observed with changes in current density at constant time of deposition. Fine close-grained structures are observed at low current as shown in Figure 1. On examination of all coated layers, some fine cracks are found in the layer deposited at higher current, that is, at 45 mA.

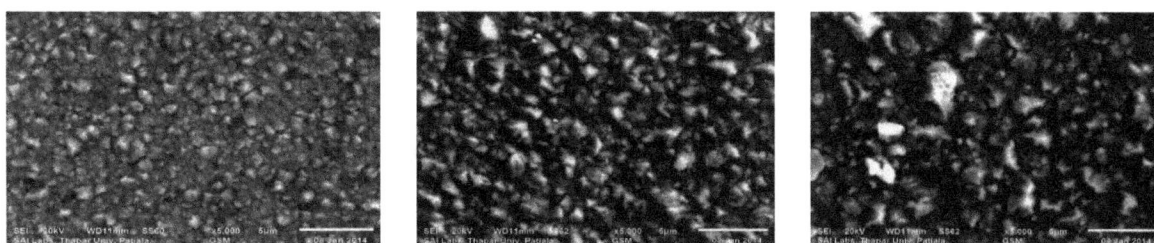

Figure 3. *SEM photographs of Ni/TiO_2 composite layers at currents of 15, 30, and 45 mA*

The microhardness of Ni-TiO_2 composite coatings is a function of the grain structure developed during the process. The increase in microhardness of Ni-TiO_2 composite coatings can be explained on the basis that TiO_2 nanoparticles are uniformly distributed in the Ni matrix. Further it is found that growth of the Ni grains, through its grain refining and dispersive strengthening effects, also affects the microhardness of coatings [13].

As shown in Fig. 3, the SEM results shows that at lower current, that is, at 15 mA, a fine close-grained structure is observed, whereas the course grain structure is observed at current intensities of 30 and 45 mA. The measurements show that the microhardness is influenced by the change in current density as shown in Fig. 4. The measurements were carried out three times for each sample to confirm the reproducibility of the results, and the maximum indentation depth was of the order of 1 μm.

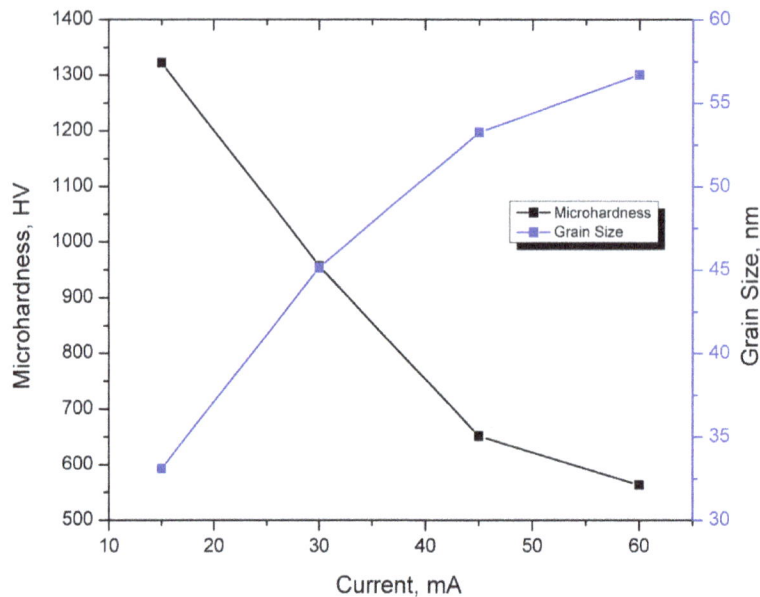

Figure 4. *Effect of current on microhardness and grain size of composite layer at pH 3.5 and temperature of 30 °C.*

In nanocomposite coatings, it is found that dispersion strengthening is largely due to the uniform distribution of TiO_2 nanoparticles in the coatings. Such particles possess higher density and fine size, and close and fine grains are obtained in the composite [14]. Codeposition of fine TiO_2 nanoparticles in Ni matrix restrained the growth of large Ni grains, resulting in the fine-grained structure of the composite layer. Further dispersive strength obtained during the reinforcement phase acts as a barrier that retards the plastic deformation of the Ni matrix, resulting in high microhardness of the composite coatings [15]. Accordingly, higher microhardness and improved friction and wear properties are found in the hardened matrix.

Effect of pH value on microhardness of Ni-TiO$_2$ coatings

In electrodeposition, the pH value of the solution is a vital parameter to be controlled to maintain the process optimization. If the pH of the Watts electrolyte is too high or too low due to chloride ion concentration, more oxygen will evolve and then hydroxyl ions will be discharged in preference to the dissolution of nickel. Under this condition, the nickel anode becomes passive and the efficiency of anode dissolution is almost zero [16,17]. The pH value of the bath influences the hydrogen evolution voltage, the precipitation of basic inclusion, and the decomposition of the complex or hydrate from which the metal is deposited. In a complex bath, the pH may influence the equilibrium between various processes. When the anode is insoluble, oxygen evolution takes place at the anode:

$$2H_2O \rightarrow O_2 + 4H^+ + 4e^-$$

On the other hand, hydrogen evolution at the cathode is accompanied by the production of hydroxide ions:

$$2H_2O + 2e^- \rightarrow 2OH^- + H_2$$

If the current efficiency is greater at the anode than at the cathode, the bath becomes more alkaline. If the electrode efficiencies are similar, the pH of the bath remains unchanged. Hence a change in pH of a plating bath is a good indication of electrode efficiencies. Since it is not possible to predict all the factors, the best pH range must be determined empirically for specific coating compositions. In this study, coating has been done at different pH concentrations of the Watts solution of 3.5, 4.5, and 5.5. The effects of the quality of coatings were investigated using constant conditions of a current density of 15 mA, temperature of 50 °C, deposition time of 1200 s, and stirring rate of 210 rpm. Figure 5 shows the microhardness of the composite layer at different pH values at a current of 15 mA.

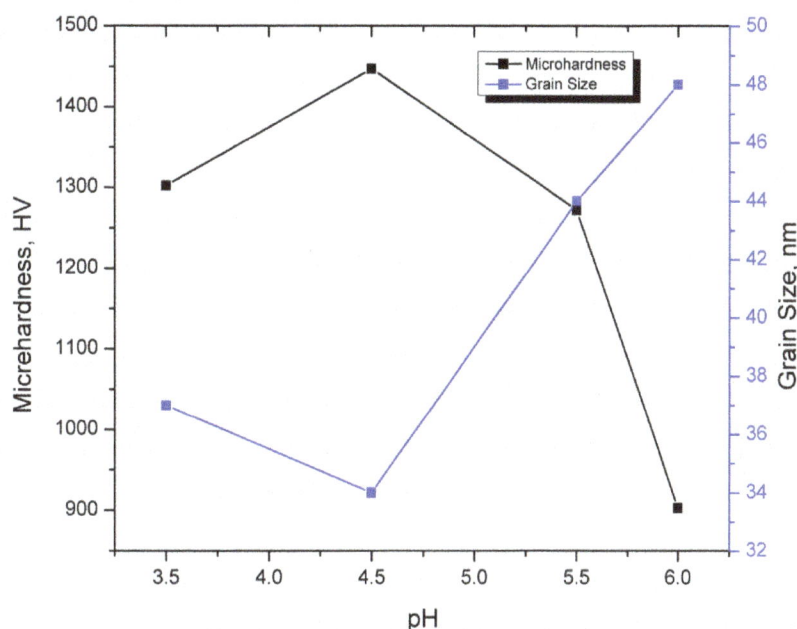

Figure 5. *Effect of pH value on microhardness and grain size of composite layer at a current of 15mA and temperature of 30 °C*

The microhardness, which is mainly dependent on the grain size, is influenced by the pH value of the solution. Figure 6 shows the SEM cross-sectional image of Ni-TiO$_2$ composite coatings produced at pH concentrations of 3.5, 4.5, and 5.5.

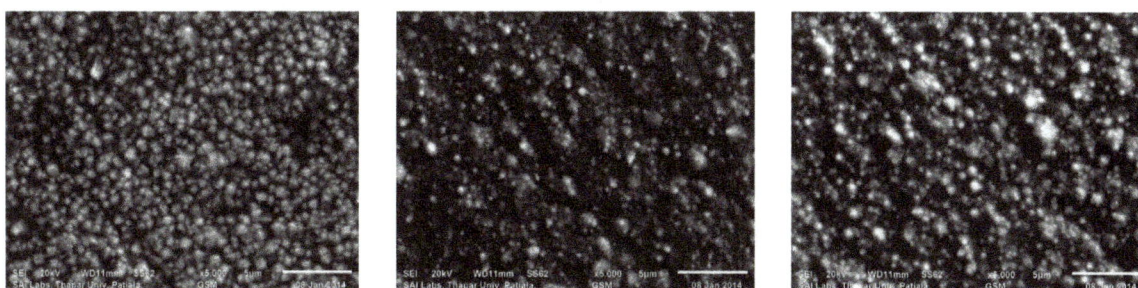

Figure 6. SEM photomicrographs of Ni/TiO$_2$composite layer at pH 4.5, 5.5, and 3.5, respectively.

The micrographs shows a dark dispersion uniformly distributed all over the coatings and XRD analysis of these dispersions reveals the dark area to be rich in TiO$_2$ particles; however the concentration of TiO$_2$ particles in the coating varied with changes in the pH of the electrolyte. The amount of TiO$_2$ particles is highest at pH 4.5.

Effect of TiO$_2$ particle concentration on the microhardness of composite coatings

The incorporation of TiO$_2$ nanoparticles in the nickel matrix leads to the formation of finer-grained structures, which further decreases the residual tensile stresses and increases the microhardness and wear and corrosion resistance of the composite coatings [18]. During electrodeposition a coating of a thin layer of one metal is deposited on top of a different metal at different concentrations of TiO$_2$ nanoparticles in the solution in order to modify the surface properties.

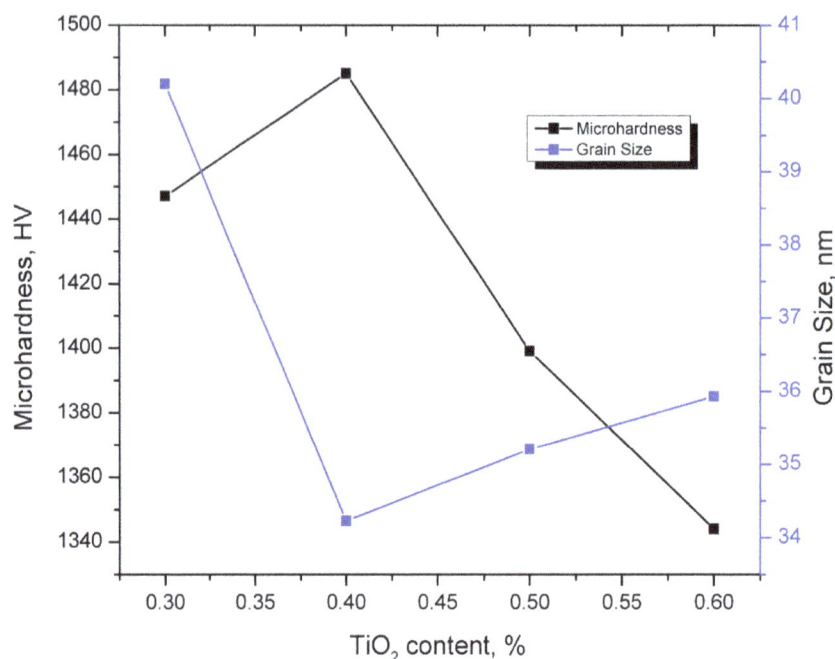

Figure 7. *Effect of TiO$_2$ content on microhardness and grain size at current = 15 mA, pH = 4.5, and temperature = 30 °C.*

Deposition was carried out by varying the concentration of Ti-O$_2$ nanoparticles in Watts solution from 0.3 to 6 g L^{-1} at a lower current, that is, 15 mA, and a solution pH of 4.5. The microhardness of each composite coating was measured and the maximum microhardness was found at 0.4 g L^{-1} solution. The addition of more solution had a negligible effect on the coating as shown in Fig. 7. When TiO$_2$ nanoparticles were added to the electrolytic bath of Watts solution, the borders of the Ni grains became more localized, resulting in a reduction in mean grain size compared to pure Ni. The addition of TiO$_2$ nanoparticles to the nickel matrix changed the structure of the composite coatings. According to Scherrer's equation, with increasing content of particles in the coating, the Ni grains decrease in size [13].

Conclusions

Electrodeposition is a versatile route for depositing composite coatings of Ni-TiO$_2$ on the surface of cutting tools. Moreover the setup cost for electrodeposition coating is quite low compared to other conventional coating processes like CVD and PVD. In this work, a Ni-TiO$_2$ nanocomposite coating was deposited on a WC cutting tool by the electrodeposition technique in order to improve the mechanical properties. The effects of the applied current, pH of the solution, and TiO$_2$ particle concentration were studied. The microstructure and phase structure of the composite layer were investigated using atomic force microscopy, scanning electronic microscopy, and X-ray diffraction.

It was found that the surface morphology of Ni-TiO$_2$ coating was crack free at lower current densities; however some fine cracks appeared at high current densities, which means that Ni-TiO$_2$ coatings are prone to cracking at higher current density. The microhardness of Ni-TiO$_2$ coatings decreases due to increases in grain size and decreases in tungsten content. The highest microhardness value is obtained at a lower current, that is, 15 mA. The pH value of the electrolyte bath influences the hydrogen evolution voltage. The microhardness of the composite layer is influenced by the pH value of the solution. A higher microhardness value is obtained at pH 4.5 at lower current density. Further addition of TiO$_2$ nanoparticles was carried out and the maximum microhardness of 1484 HV was found at 0.5 % TiO$_2$, pH 4.5, and a current of 15 mA.

References

[1] C. S. Lin, C. Y. Lee, C. F. Chang, *Surf. Coat. Technol.* **200** (2006) 3690-3697.

[2] M. Schlesinger, M. Paunovic, *Modern Electroplating*, John Wiley & Sons, Inc., Hoboken, New Jersey, USA, 2010.

[3] J. H. W. Loffler, *Surf. Coat. Technol.* **68/69** (1994) 729-736.

[4] H.-J.Schröder, G.K.Wolf, *Surf. Coat. Technol.* **74/75** (1995) 178-186.

[5] H. Kupfer, F. Richter, S. Friedrich, H-J. Spies, *Surf. Coat. Technol.* **74/75** (1995) 333-339.

[6] S. B. Sant, K. S. Gill, *Surf. Coat. Technol.* **68/69** (1994) 152-161.

[7] G. Berg, Ch. Friedrich, E. Broszeit, Ch. Berger, *Surf. Coat. Technol.* **86/87** (1996) 184-191.

[8] T. Hurkmans, D. B. Lewis, J.S. Brooks, *Surf. Coat. Technol.* **86/87** (1996) 192-201.

[9] O. Knotek, W. Bosch, M. Atzor, D. Hoffmann, J. Goebel, *High Temp.–High Press* **18** (1986) 435-443.

[10] S. Spanou, E. A. Pavlatou, *J. Appl. Electrochem.* **40** (2010) 1325-1336.

[11] W. Chen, Y. He, W. Gaoa, *J. Electrochem. Soc.* **157/8** (2010) 122-128.

[12] G. A. Di Bari, *Electrodeposition of Nickel*, in *Modern Electroplating*, John Wiley & Sons, Inc., Hoboken, NJ, USA, 2010

[13] A. Sadeghi, R. Khosroshahi, Z. Sadeghian, *J. Surf. Invest.-X-ray+* **5(1)** (2011) 186–192.

[14] L. Chen, L. Wang, Z. Zeng, J. Zhang, *Mater. Sci. Eng., A* **434** (2006) 319-328.

[15] M. Momenzadeh, S. Sanjabi, *Mater. Corros.* **63(7)** (2012) 614-619.

[16] J. Tientong, C. R. Thurber, N. D'Souza, A. Mohamed, T. D. Golden, *Int. J. Electrochem* (2013) Article ID 85386, 1-8.

[17] Z. Abdel Hamid, S.M. El-Sheikh, *J. Metall. Eng.* **2** (2013) 71-79.

[18] S. Spanou, E. A. Pavlatou, N. Spyrellis, *Proceedings of the 7th International Conference Coatings in Manufacturing Engineering*, Chalkidiki, Greece, 2008, p. 57-71.

6

Inhibition of mild steel corrosion using *Jatropha Curcas* leaf extract

OLORUNFEMI MICHAEL AJAYI⊠, JAMIU KOLAWOLE ODUSOTE*,
RAHEEM ABOLORE YAHYA*

Department of Mechanical Engineering, University of Ilorin, Ilorin, Nigeria

*Department of Materials and Metallurgical Engineering, University of Ilorin, Ilorin, Nigeria

⊠Corresponding Author

Abstract

Jatropha Curcas leaf was investigated as a green inhibitor on the degradation of mild steel in 4 M HCl and 4 M H$_2$SO$_4$ aqueous solutions using gasometric technique. Mild steel coupons of dimension 2 × 1.5 cm were immersed in test solutions of uninhibited acid and also those with extract concentrations of 4 ml, 6 ml, 8 ml and 10 ml at 30 °C, for up to 30 minutes. The results showed that as the concentration of the extract increases, there was reduction in the corrosion rate. As the extract concentration increased from 4 ml to 10 ml at 30 minutes exposure, the volume of hydrogen gas evolved decreased from 19.1 cm^3 to 11.2 cm^3 in H$_2$SO$_4$ medium, while it reduced to 5 cm^3 from 9 cm^3 in HCl medium. Also, the metal surface-phytoconstituent interaction mechanism showed that 6 minutes is the best exposure time for the adsorption of the extract in both acidic media. The Jatropha Curcas leaf extract was adsorbed on the mild steel surface to inhibit corrosion, while the experimental data obtained at 30 minutes exposure in both acidic media were well fitted with the Langmuir adsorption isotherm. Hence, Jatropha Curcas leaf extract is a good and safe inhibitor in both acidic solutions.

Keywords

Gasometric, Inhibitor, adsorption, mild steel, Langmuir isotherm

Introduction

Mild steel is a material commonly used in industries due to its low cost, availability and excellent mechanical properties [1]. However, the major drawback to its application is corrosion attack, which usually leads to structures degradation, equipment shutdown, loss of machines efficiency, and loss of valuable products, to mention but few [2]. The average corrosion cost has been reported to be about 3.5-4.5 % of the Gross National Product of most industrialized nations [3].

Corrosion can be prevented in several ways but the use of inhibitors is one of the most acceptable practices. The use of synthetic inhibitors has been seriously discouraged due to its high cost, non-biodegradability and harmfulness. Hence, naturally occurring compounds from plants origin have been a subject of interest for researchers because of their abundant availability, cost effectiveness and environmentally friendly [4]. Several studies have been carried out on the use of these naturally occurring compounds as corrosion inhibitors for metals in different media [5-19].

Jatropha Curcas (JC) is a perennial, multi-purpose and drought resistant plant that belongs to the family of *Euphorbiaceous* JC is also a tropical plant that can be grown in low to high rainfall regions [20], on both fertile and even in less fertile soil. *Jatropha* oil, obtained by crushing the seeds is used as biodiesel fuel. The plant is planted by farmers all over the world, because it is not browsed by animals. Non-toxic variety of *Jatropha* could be a potential source of oil for human consumption and the seed cake can be a protein source for humans as well as for livestock [21]. Another potential application of the leaves as corrosion inhibitor for mild steel in acidic media is established in this study.

Experimental Procedure

The chemical composition of the mild steel specimen used for this experiment in wt % is 0.17 % C, 0.21 % Si, 0.55 % Mn, 0.02 % P, 0.02 % S, 0.18 % Cu, 0.01 % Ni, 0.02 % Sn and 98.81 % Fe. Specimens were press cut into pieces with dimension of 1.5 × 2 cm coupons. The specimens were polished using LINN MAJOR STRUER-ITALY (Model No. 224732) with emery papers 140/0304 – 140/0308 grades. Subsequently, they were degreased in ethanol, dried in acetone and stored in desiccators. The solutions of HCl and H_2SO_4 were prepared by using double distilled water. The fresh leaf of *Jatropha Curcas* (JC) plant was taken, washed under running water, cut into pieces, air dried and then grounded well and sieves into powdery form. Then, 10 g each of the powdery leaf was put into flat bottom flask containing 200 cm^3 of 4 M HCl and H_2SO_4 aqueous solutions. This concentration was used in order to fasten the rate of reaction between the metal surface and the acidic extract of the inhibitor within the period of the experiment. The resulting solutions were refluxed for 2 hours and left overnight before it was carefully filtered. The stock solution was prepared from the filtrate and into the desired concentrations. In this study, extract amount of 4-10 ml correspond to 0.2 g dm^{-3}, 0.3 g dm^{-3}, 0.4 g dm^{-3} and 0.5 g dm^{-3}, respectively.

The gasometric assembly used for the measurement of hydrogen evolution was as reported by Aisha *et al.* [22]. A reaction vessel was connected to a burette through a delivery tube. The 4 M HCl solution was introduced into the mylius cell, and the initial volume of air in the burette was recorded. Then, mild steel coupon was dropped into the HCl solution, and the mylius cell was quickly closed. The volume of hydrogen gas evolved from the corrosion reaction was monitored by the volume change in the level of water in the burette. The change in volume was recorded every 120 seconds for up to 30 minutes. Similar procedure was repeated with the inhibitor. The same experimental procedure was followed for 4 M H_2SO_4 solution.

The inhibition efficiency and the degree of surface coverage were determined using Equations 1 and 2 [21]:

$$\text{Inhibition efficiency (I.E.), \%} = \frac{V_{HO} - V_{H1}}{V_{HO}} \times 100 \tag{1}$$

$$\text{Surface coverage - } \theta = \frac{V_{HO} - V_{H1}}{V_{HO}} \tag{2}$$

where V_{H0} is the volume of H_2 gas evolved without inhibitor and V_{H1} is the volume of H_2 gas evolved with inhibitor.

Results and discussion

Figure 1 shows the variation of volume of hydrogen gas evolved with time for the corrosion of mild steel in various concentrations of the inhibitor in HCl aqueous solution. As shown in the figure, the hydrogen gas was not evolved in the first 8 minutes due to slow rate of corrosion reaction at the initial stage resulting from the inability of the acidic extract to quickly penetrate the metal surface. Above this exposure time, the volume of hydrogen gas evolved increased with increasing period of exposure, but decreases with increasing concentration of acidic extract of *Jatropha Curcas* leaf. The volume of hydrogen gas evolved at 30 minutes was 21.8 cm^3 for the blank solution, while that of 4 ml, 6 ml, 8 ml and 10 ml concentrations of *Jatropha Curcas* leaf extract are 9.0, 8.0, 6.8 and 5.0 cm^3, respectively. This shows that oxide film developed faster on the surface of mild steel coupon with higher inhibitor concentration, and thus reduces the corrosion rate. The blank system having no inhibitor gave the highest hydrogen gas evolution and is far apart when compared to when varying concentration of the extract of *Jatropha Curcas* leaf was added. This may be due to the absence of inhibitor that will prevent acidic solution from reaching the metal surface [22]. Presence of oxide film causes the rate of hydrogen gas evolution to decrease (*i.e.* decrease in the rate of corrosion) [22]. Aisha *et al.* [21] also opined that increase in hydrogen evolution gas in the blank system may be due to direct reaction between the acid and the metal, since there is no adsorption layer to inhibit the reaction. Hence, the rate of hydrogen gas evolution, that is, the corrosion rate will be faster in the blank solution as compared with the inhibited. Ulaeto *et al.* [11] found that the leaf and root extracts of *Eichornia Crassipe* effectively inhibited the corrosion of mild steel in 5 M HCl, and that the extracts performed better at higher concentration.

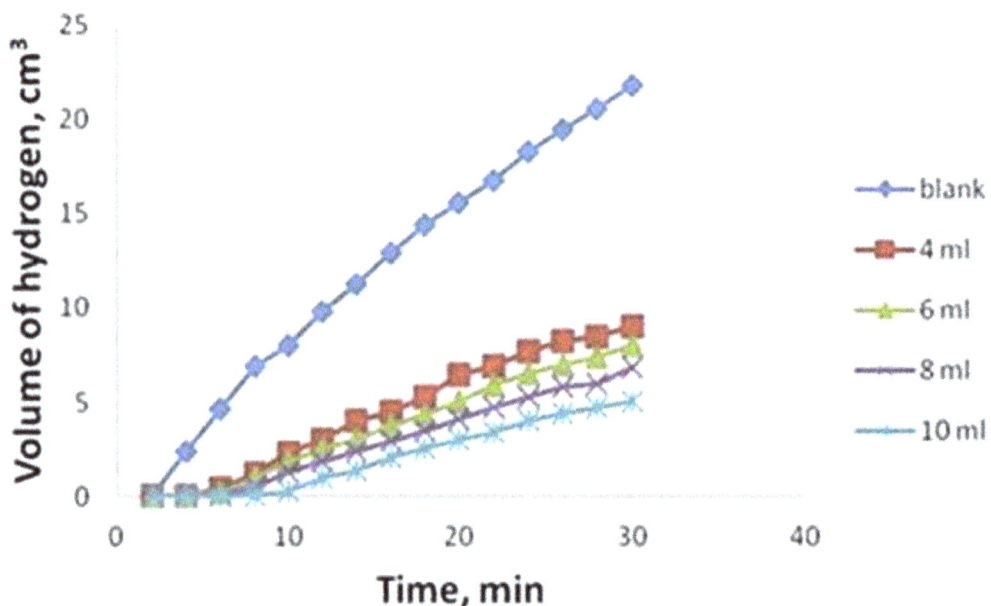

Figure 1. *Variation of volume of H₂ evolved with time of mild steel coupons for different volumes of JC extract in 4 M HCl solution*

The variation of inhibition efficiency against time of immersion with varying concentration of the inhibitor in 4 M HCl aqueous solution is shown in Figure 2. The results show that from 0 to

4 minutes, the inhibition efficiency was 0 %, corresponding to the latency period [23]. The corrosion rate was faster at the initial stage above 4 minutes, resulting in higher inhibition efficiency. However, at 6 minutes there was a re-ordering of the inhibition efficiencies from highest to the least value in descending order of the inhibitor concentration *i.e.* (10 ml < 8 ml < 6 ml < 4 ml) at all the exposure times. This revealed that there is an adsorption of the constituents of the *Jatropha Curcas* leaves extract on the surface of mild steel with 10 ml concentration of the inhibitor having the highest inhibition efficiency. The adsorption of the constituents resulted in the steady rate of corrosion (Fig. 2) due to the formation of oxide film separating the metal surface from the corrosive medium. Aisha *et al.* [21] investigated the use of *Plectranthus Tenuifloros* (Sahara) plant as safe and green inhibitor of mild steel corrosion in acidic solutions and observed that as the concentration of the extract increases, the inhibition efficiency increases. This was reported to be due to the adsorption layer formed on the surface of mild steel which inhibits the rate of corrosion. It was reported by Kuznetsov [23] that the longer the latency period, the higher the inhibition efficiency.

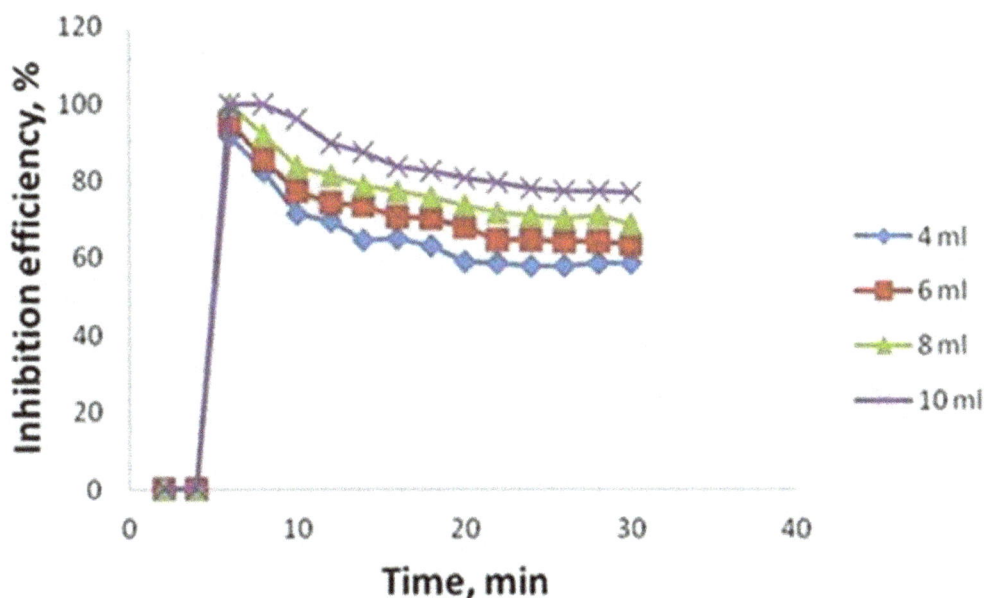

Figure 2. *Variation of inhibition efficiency with the time of immersion in 4 M HCl.*

Figure 3 shows the variation of the volume of hydrogen gas evolved with time of exposure in sulphuric acid solution. The results revealed that the corrosion rate of mild steel as indicated by the amount of H_2 gas evolved decreased in the presence of *Jatropha Curcas* leaf extract when compared to the control. The volume of hydrogen gas for blank solution was the highest as compared to those with different concentrations of *Jatropha Curcas* leaf extract. This infers that the JC leaf extract in the solution had a retarding effect on the corrosion of mild steel in H_2SO_4. Thus, the degree of inhibition can be said to be governed by the amount of JC extract present. The 10 ml concentration of the inhibitor was able to reduce the rate of hydrogen gas evolution further due to the formation of more adsorption layer on the surface of mild steel sample. The trend agrees with the result of Eddy *et al.* [10] during the determination of the inhibition efficiency of ethanol extract of *Phyllanthus Amarus* on corrosion of mild steel in H_2SO_4 solution. They reported that the volume of hydrogen decreased as the concentrations of *Phyllanthus Amarus* increased and the highest concentration of 0.5 g/L gave the least value of hydrogen gas evolution.

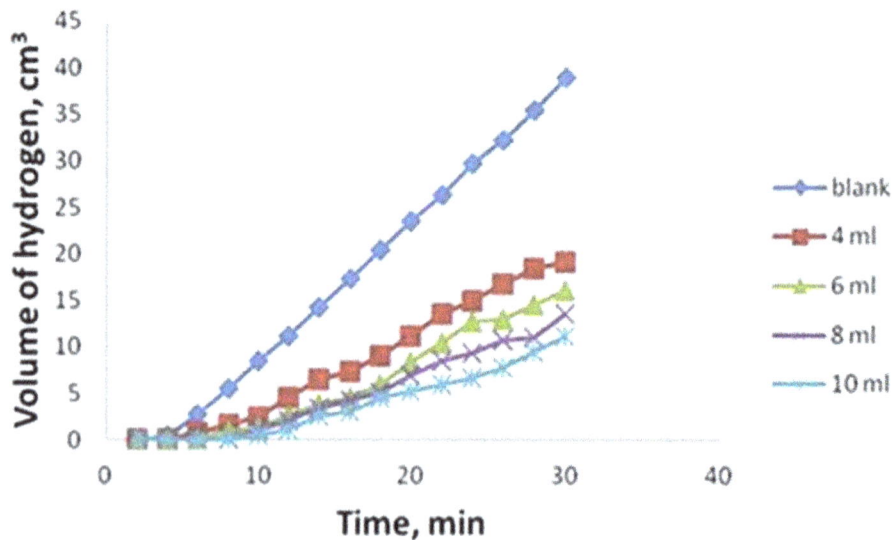

Figure 3. Variation of volume of H_2 evolved with time of mild steel in coupons for different volumes of JC extract in 4 M H_2SO_4.

Figure 4 shows the variation of percentage inhibition efficiency with time of immersion in sulphuric acid solution. The results obtained revealed that the inhibition efficiency increases as the concentration of *Jatropha Curcas* leaf extract increases, which follow similar trends with results in HCl medium. Similar observations were made when gravimetric method was used [24]. Also, from 6 minutes there was a re-ordering of the inhibition efficiencies from the highest to the least value in descending order of the inhibitor (10 ml < 8 ml < 6 ml < 4 ml) for the time of immersion. The corrosion inhibition of the plant extract on the surface of mild steel may be due to presence of the phytochemical constituents such as 1.610 mg/L alkaloid, 0.672 mg/L flavonoid, 0.412 mg/L saponins, 0.124 mg/L tannins and 0.465 mg/L phenol in the extract [24]. From the plot, it can be seen that the inhibition efficiencies of varying concentrations begin to reduce after reaching the maximum at the initial or transient stage up to 6 minutes. This stage was preceded by the latency period [23]. This reduction may be due to the faster rate of corrosion resulting from breakaway of the oxide film formed from extract inhibitor adsorption into the metal surface. However, above 20 minutes exposure at all the inhibitor concentrations, the efficiencies become relatively steady due to the protective nature of the barrier film separating the metal surface from the acidic medium.

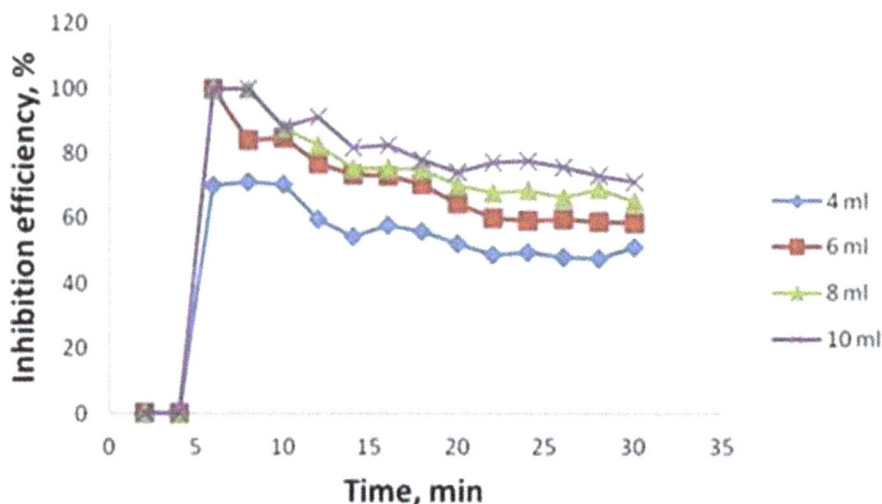

Figure 4. Variation of inhibition efficiency with the time of immersion in 4 M H_2SO_4.

Adsorption Isotherm

Adsorption isotherms are very important in knowing the mechanism of inhibition of corrosion reaction of metals. The most frequently used adsorption isotherms are Frumkin, Temkin, Freundlich, Flory Huggins and Langmuir Isotherms. However, only Langmuir Isotherms is reported in the present study, while other adsorption methods were evaluated and reported elsewhere [26,27]. Langmuir gives an expression for the concentration to the degree of surface coverage (θ) according to Equation 3 [28]:

$$C / \theta = 1 / K_{ads} + C \qquad\qquad (3)$$

Figures 5 and 6 represent the Langmuir isotherm plots of *Jatropha Curcas* leaves extract in both HCl and and H_2SO_4 aqueous solutions, respectively, showing the variation of C/θ against C at 30 minutes exposure. The plots showed that Langmuir adsorption isotherm model is appropriate for the determination of the adsorption mechanism of the extract of *Jatropha Curcas* leaves in both acidic media, since the points were well fitted linearly (as indicated by the values of coefficient of correlation, R^2, as given in Table 1) at a fixed slope of 1 according to Equation 3. The equilibrium constant of adsorption isotherm, K_{ads}, of the *Jatropha Curcas* leaf extract in both HCl and H_2SO_4 media were obtained from the intercept and the results are presented in Table 1. However, due to the complexity of the compounds in the extracts of leaves of *J. curcas* [29], it is not possible to determine the exact molecular weight of the inhibitor and hence the concentration in mol dm^{-3}. As a result, values of the standard free energy of adsorption (ΔG_{ads}) in both media could not be calculated [30].

Table 1. *Calculated values of Langmuir adsorption isotherm parameters of* Jatropha Curcas *extract in HCl and H_2SO_4 aqueous solution at 30 minutes*

Plant Extract	Concentration on intercept point, g dm^{-3}	Slope	K_{ads} / g dm^{-3}	R^2
JC in 4M HCl	0.177	1.000	5.65	0.999
JC in 4M H_2SO_4	0.206	1.000	4.85	0.995

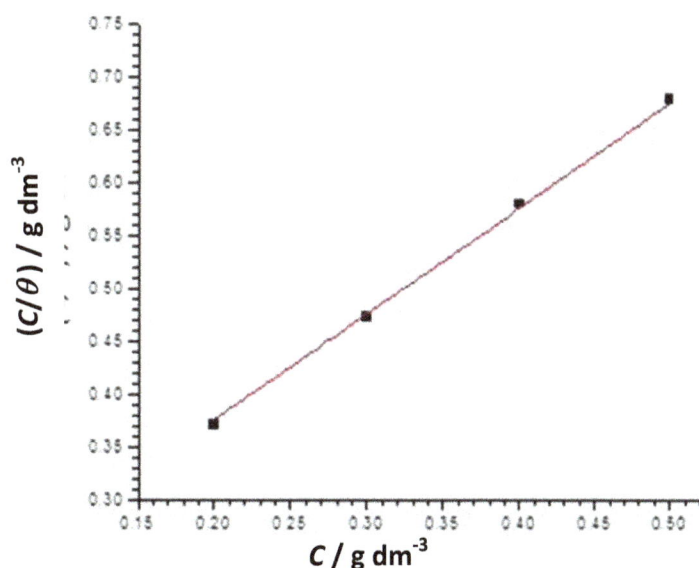

Figure 5. *Langmuir isotherm for the adsorption of the extract of* Jatropha *leaves on the surface of mild steel in 4 M HCl at 30 minutes exposure.*

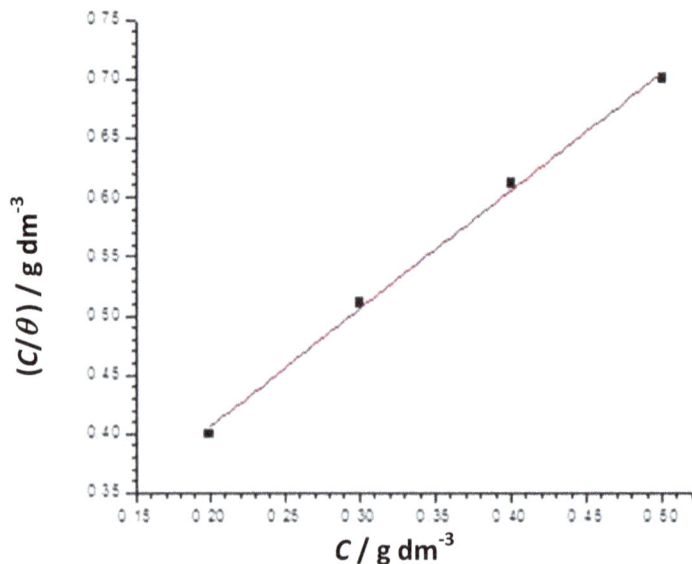

Figure 6. *Langmuir isotherm for the adsorption of the extract of* Jatropha leaves *on the surface of mild steel in 4 M H₂SO₄ at 30 minutes exposure.*

Plant extract contains organic compounds having polar atoms or groups which are adsorbed on the metal surface. Obot and Obi-Egbedi [13] reported that compounds interact by mutual repulsion or attraction when *Ipomoea Involcrata* plant extract was used as an inhibitor. This may be advocated as the reason for the slight departure of the slope values from unity as explained by Obot and Obi-Egbedi [13]. Although, in this study, the slope is fixed at 1 prior to linear fitting but few points were still slightly deviated from the straight line, which may be due to mutual repulsion or attraction of the polar atoms or groups as observed by Obot and Obi-Egbedi [13].

In addition, the adsorption of the *Jatropha Curcas* leaves extract on the mild steel surface may not involve the interaction of the adsorbate molecules with one another. According to Nnanna *et al.* [7], it was assumed that there was no interaction between the adsorbate molecules in the derivation of Langmuir isotherm. The adsorption was also assumed to be monolayer because the sites on the metal surface were taken to be energetically identical and uniformly distributed [8]. However, the adsorption process may be assumed to be due to an electrostatic interaction between the polar atoms/ions on the metal surface and the adsorbate molecules [7,29].

Conclusions

- The leaf extract of *Jatropha Curcas* acts as a good and efficient inhibitor for corrosion of mild steel in HCl and H₂SO₄ solutions.
- Inhibition efficiencies of the *Jatropha Curcas* leaf extract in HCl medium were higher than those in H₂SO₄ environment. After 30 minutes exposure with extract concentration of 10 ml, the efficiency is 77.1 % in HCl medium while 71.3 % was obtained in H₂SO₄ medium.
- The inhibition of the corrosion of mild steel by acid extract of JC is due to the phytochemical constituents in the plant extract.
- The experimental data obtained at 30 minutes exposure in both HCl and H₂SO₄ solutions with *Jatropha Curcas* leaf extract were well fitted with the Langmuir adsorption isotherm indicating that the Langmuir adsorption model is applicable in the corrosion inhibition mechanism.
- Further work will be carried out using other techniques with micrographs from SEM to show the effect of temperature and/or pH on the corrosion efficiency of *Jatropha Curcas* leaf extract on mild steel and other materials.

References

[1] J. R. Vimala, A. L. Rose, S. Raja, *Int. J. ChemTech Res.* **3** (2011) 1791-1801.

[2] P. B. Raja, M. G. Sethuraman, *J. Pigment. Resin Technol.* **38** (2009) 33-37.

[3] Corrosion costs and Preventive Strategies in the United States, http://www.cctechnologies.com (accessed 20th January 2013).

[4] O. A. Omotosho, O. O. Ajayi, V. O. Ifepe, *J. Mater. Environ. Sci.* **2** (2011) 186-195.

[5] D. P. Rani, S. Selvaraj, *Arch. App. Sci. Res.* **2(6)** (2010) 140-150.

[6] N. S. Patel, S. Jauhari, G.N. Mehta, *Arab. J. Sci. Eng.* **34** (2009) 61-69.

[7] L. A. Nnanna, V.U. Obasi, O. C. Nwadiuko, K. I. Mejeh, N. D. Ekekwe, S. C. Udensi, *Arch. App. Sci. Res.* **4** (2012) 207-217.

[8] O. M. Ndibe, M.C. Menkiti, M. N. C. Ijomah, O. D. Onukwuli, *Electron. J. Environ. Agric. Food Chem.* **10** (2011) 2847-2860.

[9] C. A. Loto, R.T. Loto, A.P.I. Popoola, *Int. J. Phys. Sci.* **6** (2011) 3689-3696.

[10] N. O. Eddy, *Port. Electrochim. Acta.* **27** (2009) 579-589.

[11] S. B. Ulaeto, U. J. Ekpe, M. A. Chidiebere, E. E. Oguzie, *Int. J. Mater. Chem.* **2** (2012) 158-164.

[12] A. M. Al-Turkustani, S. T. *Arab. Int. J. Chem.* **2** (2010) 54-76.

[13] I. B. Obot, S. A. Umoren, N. O. Obi-Egbedi, *J. Mat. Environ. Sci.* **2** (2011) 60-71.

[14] C. A. Loto, *J. Mater. Environ. Sci.* **2** (2011) 335-344.

[15] K. P. V. Kumar, M.S.N. Pillai, G. R. Thusnavis, *J. Mater. Sci. Technol.* **27** (2011) 1143-1149.

[16] O. O. Ajayi, O. A. Omotosho, K. O. Ajanaku, O. O. Babatunde, *J. Eng. App. Sci.* **6** (2011) 10-17.

[17] R. A. L. Sathiyanathan, S. Maruthamuthu, M. Selvanayagam, S. Mohanan, N. Palaniswamy, *Int. J. Chem. Tech.* **12** (2005) 356-360.

[18] R. Chauhan, U. Garg, R.K. Tak, *E-Journal of Chemistry* **8** (2011) 85-90.

[19] J. T. Nwabanne, V. N. Okafor, *J. Emerg. Trends Eng. Appl. Sci.* **2** (2011) 619-625.

[20] L. Jatropha Curcas, http://www.dovebiotech.com (accessed on 15th March 2013).

[21] M. A. Aisha, M. A. Nabeeh, *Global Juornal of Science and Frontier Research Chemistry* **12** (2012) 73-84.

[22] A. M. Al-Turkustani, S. T. Arab, R. H. Dahiri, *Modern Applied Science* **4** (2010) 105-124.

[23] Y. I. Kuznetsov, *Russ. Chem. Rev.* **73** (2004) 75-87.

[24] J. K. Odusote, O. M. Ajayi, *J. Electrochem. Sci. Technol.* **4** (2013) 81-87.

[25] O. M. Ajayi, *M.Sc. Project Report*, Department of Mechanical Engineering, University of Ilorin, Ilorin, Nigeria, 2013.

[26] N.O. Eddy, S.A. Odoemelam, A. J. Mbaba, *Afri. J. Pure Appl. Chem.* **2** (2008) 132-138.

[27] R. Staubmann, M. Schubert-Zsilavecz, A. Hiermann and T. Kartnig, *Phytochemistry,* **50** (1999) 337-338.

[28] S. Rekkab, H. Zarrok, R. Salghi, A. Zarrok, Lh. Bazzi, B. Hammouti, Z. Kabouche, R. Touzani, M. Zougagh, *J. Mater. Environ. Sci.* **3**(2012) 613-627.

[29] E. E. Ebenso, N. O. Eddy and A. O. Odiongenyi, *Afr. J. Pure Applied Chem.* **2** (2008) 107-115.

7

Cerium oxide as conversion coating for the corrosion protection of aluminum

JELENA GULICOVSKI✉, JELENA BAJAT*, VESNA MIŠKOVIĆ-STANKOVIĆ*,
BOJAN JOKIĆ*, VLADIMIR PANIĆ**, SLOBODAN MILONJIĆ

Vinča Institute of Nuclear Sciences, University of Belgrade, P. O. Box 522, 11000 Belgrade, Serbia
Faculty of Technology and Metallurgy, University of Belgrade, Karnegijeva 4, 11120 Belgrade, Serbia
**ICTM – Center of Electrochemistry, University of Belgrade, Njegoševa 12, 11000 Belgrade, Serbia*

✉Corresponding Author

Abstract

CeO_2 coatings were formed on the aluminum after Al surface preparation, by dripping the ceria sol, previously prepared by forced hydrolysis of $Ce(NO_3)_4$. The anticorrosive properties of ceria coatings were investigated by the electrochemical impedance spectroscopy (EIS) during the exposure to 0.03% NaCl. The morphology of the coatings was examined by the scanning electron microscopy (SEM). EIS data indicated considerably larger corrosion resistance of CeO_2-coated aluminum than for bare Al. The corrosion processes on Al below CeO_2 coating are subjected to more pronounced diffusion limitations in comparison to the processes below passive aluminum oxide film, as the consequence of the formation of highly compact protective coating. The results show that the deposition of ceria coatings is an effective way to improve corrosion resistance for aluminum.

Keywords
Ceria, Coatings, Corrosion, Microstructure.

Introduction

It is well known that chromate-based chemical conversions were used for corrosion protection of aluminum for decades. On the other hand, because of toxic and carcinogenic properties of chromium, this kind of coatings should be soon withdrawn from all industrial processes [1,2].

Many researchers tried to find alternative for chromate conversions; it is believed that rare earth elements (cerium, lanthanum, neodymium and yttrium) could effectively inhibit the corrosion of aluminum. Hinton *et al.* [3,4] introduced the application of lanthanide compounds for this purpose. The reported results showed that these barrier coatings simply prevent contact between aluminum surface and corrosive species. The best degree of corrosion inhibition is achieved with cerium [5-8].

Based on these findings, the aim of this study was to examine efficacy of ceria based coating for corrosion protection of aluminum, by means of the scanning electron microscopy (SEM) and the electrochemical impedance spectroscopy (EIS).

Experimental

Materials

The aluminum panels, 20 mm×18 mm×1 mm in size, were used as substrates. The surface of specimens was mechanically polished by 800, 1200, 2400 and 4000 grit emery papers. To remove the surface contamination the panels were degreased in alkaline solution containing: NaOH, $Na_3PO_4 \cdot 12H_2O$, Na_2SiO_3 and ethoxylate of nonylphenol with nine molecules of ethylene oxide, during 1 min at 70 °C. Finally, the panels were rinsed by distilled water and dried in air at room temperature.

For the investigation in this study ceria sol which was obtained by forced hydrolysis of 1 N $Ce(NO_3)_4$ aqueous solution was used. The ceria sol was synthesized by Gulicovski *et al.* [9] according to the reported procedure. The main characteristics of the synthesized ceria sol are: pH value of synthesized sol, 0.66, average particle size, 71 nm, and solid phase content, 0.51 mass%. In order to obtain the stable and pure ceria dispersion, nitrates were removed from the prepared sol by ultrafiltration. The final pH of the sol was around 4.

The CeO_2 coatings were formed on the aluminum panels by dripping method. The ceria sol was applied in the quantity of 2×30 μl. Between the two applications, dispersing medium was evaporated at 35 °C.

Methods

The corrosion behavior of the ceria-coated aluminum was determined by the electrochemical impedance spectroscopy (EIS). EIS data were recorded at the open circuit potential using a Gamry, Reference 600 potentiostat/galvanostat/ZRA.

To examine the surface morphology of the coatings, scanning electron microscope (SEM) JOEL, model JSM-5800, at 20 kV was used.

Results and Discussion

Surface morphology of bare Al and Al/CeO₂

Figure 1 shows typical SEM images of the Al and Al/CeO₂ specimens before and after prolonged exposure to 0.03% NaCl solution. An uniform pattern of the Al surface can be seen after mechanical polishing and degreasing (Figs. 1a and 1b). A highly porous film consisted of uniform μm-sized grains is observed at the Al surface after immersion in NaCl solution (Figs. 1c and 1d), *i.e.*, native homogeneous surface has been destroyed and a new layer of corrosion products was produced.

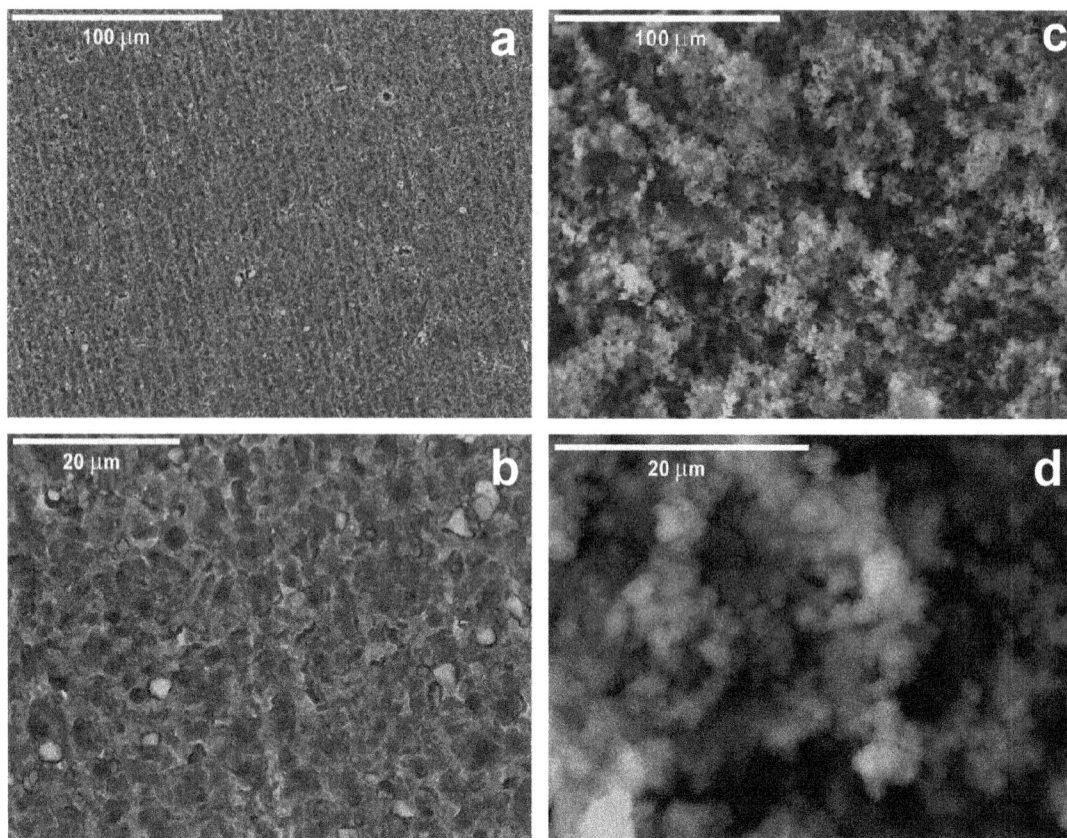

Figure 1. *Typical scanning electron micrographs of Al sample at two different magnifications: before (a and b) and after exposure to 0.03% NaCl (c and d).*

Typical surface morphology of a cerium oxide coating deposited onto Al panel is shown in Fig. 2. Narrow cracks and sparing pits are visible (Figs. 2a and 2b); however, no signs of the peeling were detected. The cracks are typical for rather thick conversion film and it is the result of the stress induced in the film during the drying process. SEM micrographs of the film surface after immersion in 0.03% NaCl solution (Figs. 2c and 2d) revealed presence of agglomerates of corrosion products (Fig. 2c). The agglomerates appear to lay over compact granular structure (Fig. 2d), uniformly consisted of μm-sized fluffy grains.

Corrosion behavior

The corrosion behavior and stability of prepared Al/CeO$_2$ sample were investigated by EIS after different times of the exposure to 0.03 % NaCl. The data registered after 24 h of exposure are shown in Fig. 3 as complex plane and Bode phase shift spectra and compared to the data of Al reference sample. The complex plane spectra are consisted of at least three semicircles. The semicircle at lowest frequencies (below 0.4 Hz) is more pronounced for Al/CeO$_2$, whereas it is only weakly indicated in the spectrum of Al (below 0.3 Hz). The peaks corresponding to two the high frequency semicircles appear clearly in the phase shift spectra, whereas low frequency semicircle can be observed only as a shoulder around 0.1 Hz for Al/CeO$_2$. However, phase shifts at low frequencies for both Al/CeO$_2$ and Al are similar, around 20°. The semicircles are larger for Al/CeO$_2$, especially two of them appearing at lower frequencies. This indicates good corrosion protection of the applied ceria coating, since this kind of the spectra, usually observed for the corrosion of aluminum, are assigned to the passive Al-oxide film (high frequency data) and corrosion-related charge transfer processes beneath (low frequency data) [10-12].

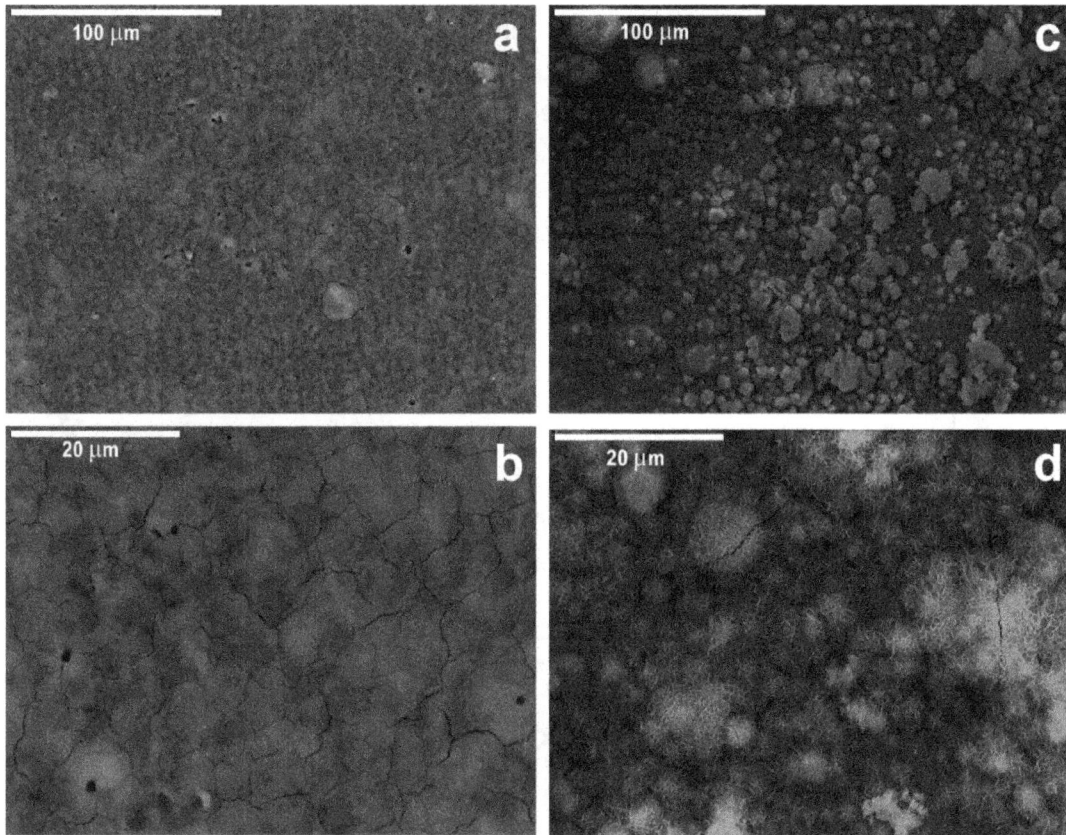

Figure 2. *Typical scanning electron macrographs of Al/CeO₂ sample at two different magnifications: before (a and b) and after exposure to 0.03% NaCl (c and d).*

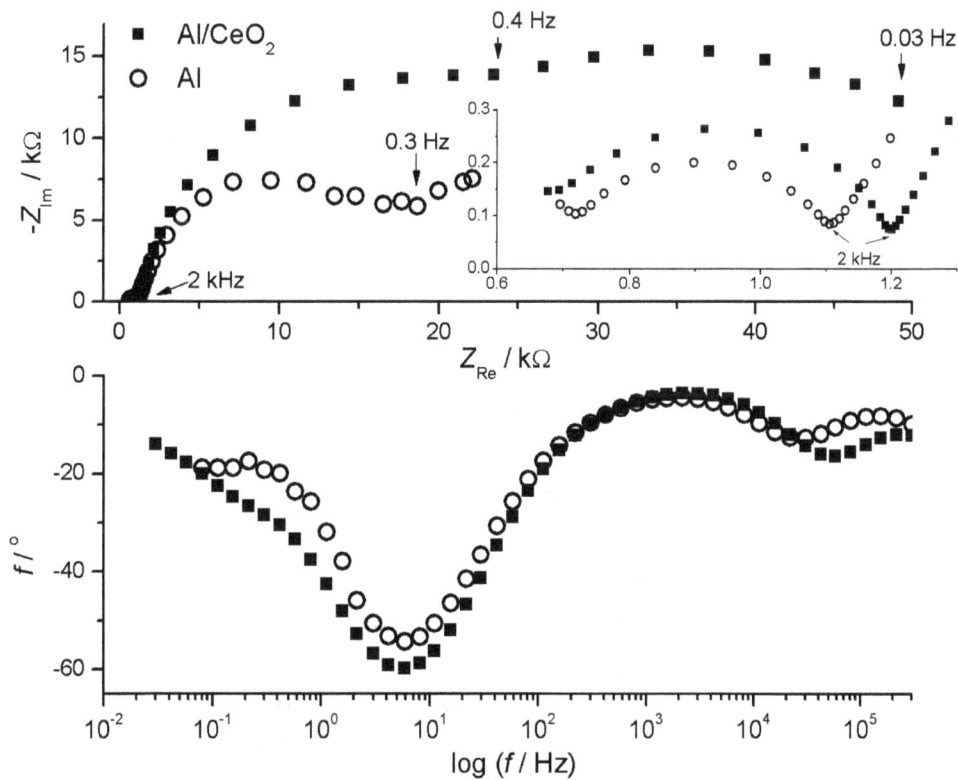

Figure 3. *Complex plane and Bode plots of Al and Al/CeO₂ sample registered at open circuit potential after 24 h exposure to 0.03 % NaCl. Inset: high frequency parts of the complex plane spectra.*

In order to examine the effect of exposure time on the characteristic features of the EIS spectra, they were recorded at prolonged exposure times. Typical complex plain and Bode plots of EIS data are presented in Fig. 4 by those registered after 312 h of exposure.

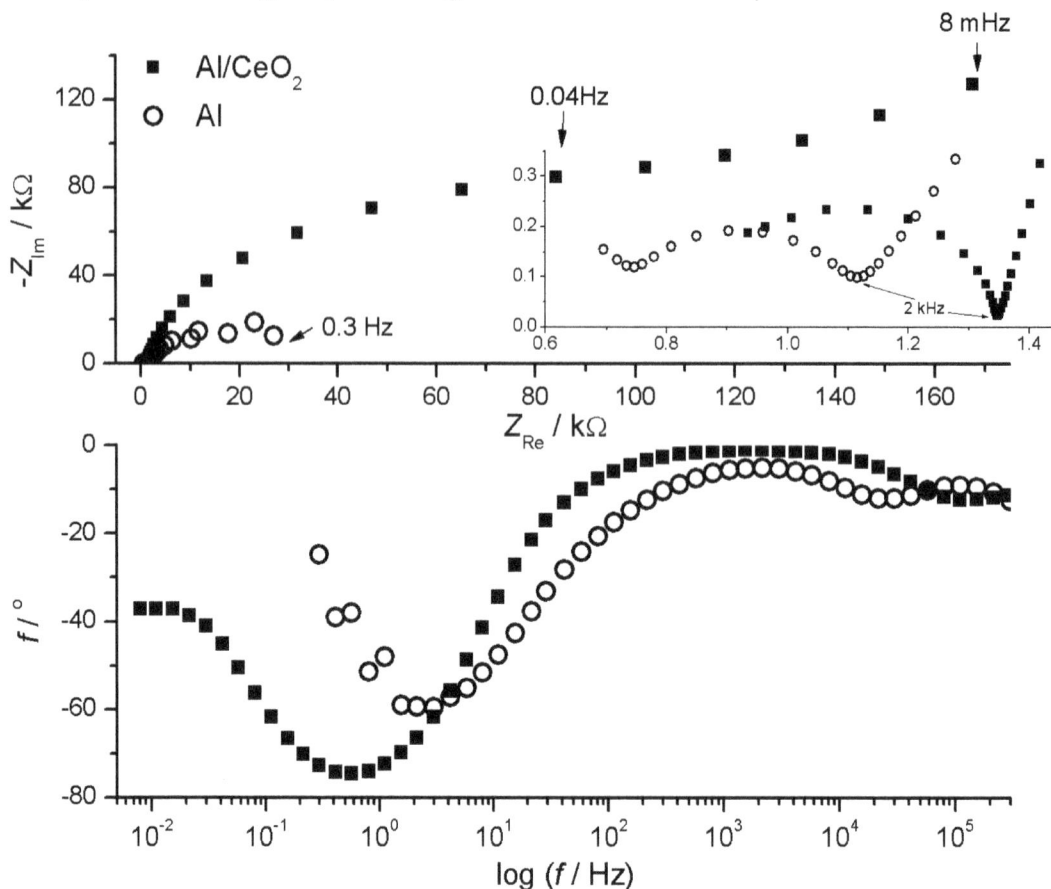

Figure 4. *Complex plane and Bode plots of Al and Al/CeO$_2$ sample registered at open circuit potential after 312 h exposure to 0.03 % NaCl. Inset: high frequency parts of the complex plane spectra.*

As Fig. 4 shows, prolonged exposure to the corrosive environment does not considerably change the characteristic features of the spectra. The comparison of Figs. 3 and 4 reveals that the spectra for both Al and Al/CeO$_2$ samples at high frequencies are poorly affected by the prolongation of the exposure, which indicates that 24 h of exposure is sufficiently long to allow complete formation of passive oxide layer. However, it appears that corrosion processes on bare Al are not hindered since its spectra from Figs. 3 and 4 at lower frequencies, related to the charge transfer, are similar. It follows that corrosion products formed after 24 h are not incorporated into the passive film with impedance response at high frequencies.

These observations, except those related to the high frequency semicircles and passive aluminum-oxide interlayer does not hold for Al/CeO$_2$ sample. It is to be noted from Figs. 3 and 4 that the semicircles for Al/CeO$_2$ sample in low frequency domain grow considerably upon the increase in exposure time, which can be considered as a result of increased corrosion resistance with respect to bare Al. At lowest frequencies (below 0.04 Hz), the semicircle in Fig. 4 is followed by a straight line assignable to the diffusion controlled corrosion processes. Indeed, the corresponding Bode plot reaches the values around 40°, which is quite close to the theoretical value of 45°. Well pronounced diffusion limitations are due to the presence of a compact protective film induced by CeO$_2$.

Conclusions

With an appropriate surface pre-treatment, ceria was formed on the aluminum panels by dripping method. These substrates were characterized by SEM and EIS methods in order to study their possible applications as protective barriers in corrosive environment.

SEM analysis confirmed formation of the oxide film on Al and Al/CeO$_2$ substrates after immersion in 0.03 % NaCl solution. The film is consisted of Al$_2$O$_3$ agglomerates with compact granular structure that improve corrosion resistance of the ceria coatings.

The results achieved by EIS measurements show that the semicircles obtained for Al/CeO$_2$ sample in low frequency domain grow considerably upon the increase in exposure in NaCl, which can be considered as a result of increased corrosion resistance with respect to bare Al.

It can be concluded that deposition of ceria coating is an effective way to improve corrosion resistance for aluminum.

Acknowledgement: This work was financially supported by the Ministry of Education, Science and Technological Development of the Republic of Serbia (Project number: 45012)

References

[1]   S.H. Abdel, *Surface Coatings Tech.* **200** (2006) 3786-3792.
[2]   J. Bieber, *Mater. Lett.* **105** (2007) 425-435.
[3]   B.R.W. Hinton, D.R. Arnott, N.E. Ryan, *Metals Forum* **7** (1984) 211-217.
[4]   B.R.W. Hinton, D.R. Arnott, N.E. Ryan, *Metals Forum* **9** (1986) 162-173.
[5]   G.F. William, M.J. O'Keefe, H. Zhou, J.T. Grant, *Surface Coatings Tech.* **155** (2002) 208-213.
[6]   A. Decroly, J.P. Petitjean, *Surface Coatings Tech.* **194** (2005) 1-9.
[7]   P. Campestrini, H. Terryn, A. Hovestad, J.H.W. de Wit, *Surface Coatings Tech.* **176** (2004) 365-381.
[8]   M.A. Dominguez-Crespo, A.M. Torres-Huerta, S.E., Rodil, E. Ramirez-Meneses, G.G. Suarez-Velazquez, M.A. Hernandez-Perez, *Electrochim. Acta* **55** (2009) 498-503.
[9]   J.J. Gulicovski, S.K., Milonjić, K. Meszaros Szecsenyi, K., *Mater. Manufacturing Proc.* **24** (2009) 1080-1085.
[10] M. Metikoš-Huković, R. Babić, Z. Grubač, *J. Appl. Electrochem.* **32** (2002) 35-41.
[11] C.M.A. Brett, *J. Appl. Electrochem.* **20** (1990)1000-1003.
[12] F.H. Cao, Z. Zhang, J.F. Li, Y.L. Cheng, J.Q. Zhang, C.N. Cao, *Mater. Corros.* **55** (2004) 18-23.

8

Effect of heat treatment on structure and properties of multilayer Zn-Ni alloy coatings

VAISHAKA R. RAO, A. CHITHARANJAN HEGDE⊠ and K. UDAYA BHAT*

Electrochemistry Research Laboratory, Department of Chemistry, National Institute of Technology Karnataka, Surathkal, Srinivasnagar-575025, India

*Department of Metallurgy and Materials Engineering, National Institute of Technology Karnataka, Surathkal, Srinivasnagar-575025, India

⊠Corresponding Author

Abstract

Composition modulated multilayer alloy (CMMA) coatings of Zn-Ni were electrodeposited galvanostatically on mild steel (MS) for enhanced corrosion protection using single bath technique. Successive layers of Zn-Ni alloys, having alternately different composition were obtained in nanometer scale by making the cathode current to cycle between two values, called cyclic cathode current densities (CCCD's). The coatings configuration, in terms of compositions and thicknesses were optimized, and their corrosion performances were evaluated in 5 % NaCl by electrochemical methods. The corrosion rates (CR)'s of multilayer alloy coatings were found to decrease drastically (35 times) with increase in number of layers (only up to 300 layers), compared to monolayer alloy deposited from the same bath. Surface study was carried with SEM, while XRD was used to determine metal lattice parameters, texture and phase composition of the coatings. The effect of heat treatment on surface morphology, thickness, hardness and corrosion behaviour of multilayer Zn-Ni alloy coatings were studied. The significant structural modification due to heat treatment is not accompanied by any decrease in corrosion rate. This effect is related to the formation of a less disordered lattice for multilayer Zn-Ni alloy coatings.

Keywords
Multilayer Zn-Ni alloy; Corrosion study; Heat treatment; SEM; XRD.

Introduction

An advanced coating technique, called composition modulated multilayer alloy (CMMA) coating is gaining interest due to their improved properties, such as mechanical strength/hardness, enhanced diffusivity, improved ductility/toughness, reduced density, reduced elastic modulus, increased specific heat, higher thermal expansion coefficient, lower thermal conductivity, enhanced corrosion and wear resistance, superior reflectance, soft magnetic properties, giant magnetoresistence and corrosion resistance not attainable in any of the metallurgical alloys [1]. The CMMA materials basically consists of alternating layers of metals/alloys on micro/nanometer scale, deposited electrolytically by making the cathode to cycle between two current densities at definite time intervals. The coating with improved resistance to highly aggressive environmental condition is demanded for the extended safe service life of industrial objects. Hence CMMA coating is well studied while finding its wide spread industrial applications [2-4].

Blum first introduced the electrodeposition of multilayered alloy on Cu-Ni, demonstrating the deposition of alternate Cu and Ni layers, tens of microns thick, from two different electrolytes [5]. The deposition using Single Bath Technique (SBT) for the fabrication of modulated alloys was recorded by Brenner [6]. Tench and White proposed that the presence of Ni in the deposit was responsible for improved corrosion resistance of fabricated Cu/Ni metal multilayers and also their mechanical properties [7]. The Zn-Ni alloy electrodeposition has attracted interest of scientific community because these alloys were found to be more corrosion resistant and thermally stable than pure zinc [8]. Many authors have attempted to understand the characteristics of this deposition process [9-12]. The abrupt change in the composition of the alloys and in the current efficiency is observed during this complex codeposition at a given value of deposition potential/current density. Multilayer coating by electrolytic method is the most promising approach for improving the corrosion resistance of Zn-Ni alloys, and is of distinct commercial interest [13-16]. The CMMA coating with Zn-Fe alloy as alternate layers was tested to exhibit better corrosion protection than individual metals. The presence of high content of Fe in Zn-Fe alloy was reasoned to be responsible for enhanced corrosion protection [17].

The thermal stability of the Zn-Ni alloy coatings is important for their general applications, especially where they are expected to perform at elevated temperature conditions, such as in some automotive applications. However, the conventional methods usually modify the physical properties of the metal being protected with the limitations, such as susceptibility to damage by heat, cost, and formation of oxide products. The real challenge to overcome these problem would be to develop a novel protection coating with an exceptional thermal stability with minimum changes to the physical properties of the protected metal. Though there are many reports with regard to the development of CMMA Zn-Ni coatings using different baths and additives, no work has been reported with regard to examine the thermal stability of those coatings [18-21]. Hence the present paper reports the development of CMMA Zn-Ni alloy coatings on mild steel (MS) from acid chloride bath using gelatin and glycerol as additives. The corrosion stability of CMMA Zn-Ni coatings on heat treatment have been tested by subjecting the coated specimens to different temperatures. The CMMA Zn-Ni coatings have been tested for their thickness, hardness, composition, surface morphology and corrosion stability before and after heat treatment, and results are discussed.

Experimental

All electrodeposition were carried out from the same electrolytic bath prepared using analytical grade reagents and double distilled water. Conventional Hull cell method was used to examine the effect of current density (c.d.) and bath constituents [22]. The Zn-Ni alloy bath was prepared by adding known amount of gelatin in hot distilled water (insoluble in cold water) and glycerol as additives, to impart brightness to the coating. Electroplating process was carried out in a stirred solution on a pre-cleaned MS panels, having 7.5 cm^2 active surface area, at 30 °C and pH 4.0. The Zn anode was used with the same exposed area. The electroplating of both monolayer (monolithic) and CMMA Zn-Ni coating was carried out using computer controlled DC power source, having output speeds of up to 160 microseconds per step voltage/current change (N6705A, Agilent Technologies) for 10 min (~15 μm thickness), for comparison purpose. The coating thickness were determined by Faraday's law and verified by measuring in Digital Thickness Meter (Coatmeasure - M&C, AA Industries/Yuyutsu Instruments). The hardness of coatings was measured using Digital Micro Hardness Tester (CLEMEX, Model: MMT-X7). The electrodeposited MS plates were subjected to heat treatment by keeping in temperature controlled oven (Technico Ind. Ltd., 3144) at temperature 100, 200, 300 and 400 °C for constant time duration of 60 minutes. The modulation in the composition of alternate layer was affected by pulsing the current periodically. The power pattern used for deposition of monolayer (direct Current) and multilayer (pulsed current) coatings is shown in Figure 1. The optimal composition and operating parameters of the bath is given in Table 1.

The corrosion behavior of the coatings were evaluated in 5 % NaCl solution at pH 4.0 using Potentiostat/Galvanostat (ACM Instruments, Gill AC Series No-1480) at temperature 25 °C, using saturated calomel electrode (SCE) as reference, and platinum as counter electrodes, respectively. The corrosion rates (CR) were measured by Tafel extrapolation method at scan rate of 1 mV s^{-1} with start potential +250 mV to reverse potential -500 mV. The electrochemical impedance spectroscopy (EIS) measurements were made in frequency range of 100 kHz to 0.01 Hz at ±10 mV perturbing voltage, to evaluate the barrier property of the coatings. The surface morphology and cross sectional view of the CMMA coating were examined under Scanning Electron Microscopy (SEM, Model JSM-6380 LA from JEOL, Japan). The variation in the phase structure of alloys in different layers were confirmed by X-ray diffraction (XRD) study, using Cu Kα ($\lambda = 0.15405$ nm) radiation, in continuous scan mode with a scan rate of 2°min^{-1} (XRD, JEOL JDX-8P). Conveniently, Zn-Ni CMMA coatings are represented as: (Zn-Ni)$_{1/2/n}$ (where 1 and 2 indicate the first and second cathode current density (CCCD's) and 'n' represent the number of layers formed during total plating time. *i.e.* 10 min.

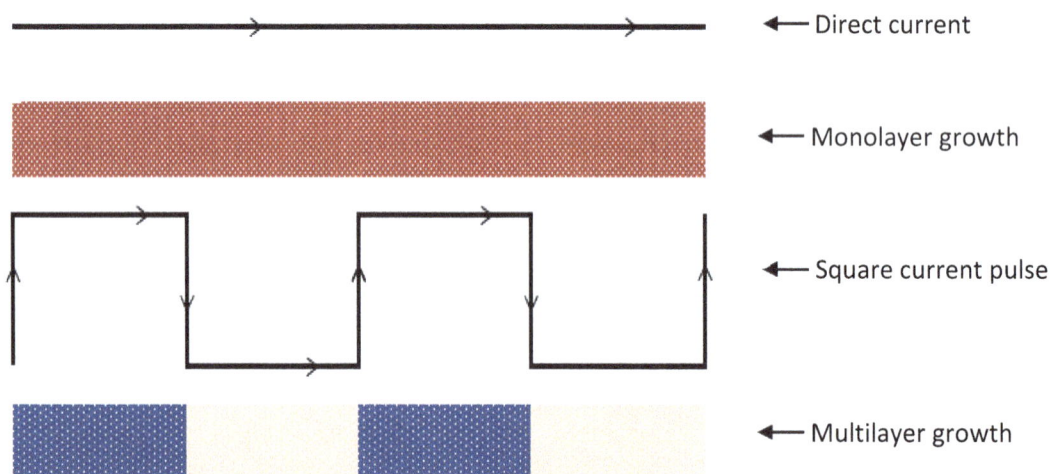

Figure 1. *Power pattern used for deposition of monolayer (direct current) and multilayer (square current pulse) coatings.*

Results and Discussion

Monolayer Zn-Ni alloy coatings

The Hull cell study confirmed that 3.0 A dm^{-2} is the optimal c.d. for deposition of monolayer Zn-Ni alloy from the bath (optimized) given in Table 1. The Zn-Ni alloy coating developed at

3.0 A dm^{-2} with ~8.0 wt.% Ni exhibited the least CR (14.46×10^{-2} mm y^{-1}) compared to other coatings at other c.d.'s as reported in Table 2.

Table 1. *Bath composition and operating parameters of the optimized bath.*

Bath ingredients	Concentration, g L^{-1}	Operating parameters
ZnCl$_2$	27.2	
NiCl$_2$ x 6H$_2$O	94.9	Anode: Pure Zinc
Boric acid	27.7	Cathode: Mild Steel
NH$_4$Cl	100	pH 4.0
Gelatin	5.0	Temperature: 30 °C
Glycerol	2.5	

From the experimental data given in Table 2, it may be noted that the wt.% Ni in the electrodeposited Zn-Ni alloy at all c.d. are less than that in the bath (64.5 % Ni). Hence it may be inferred that the proposed bath follow anomalous type of codeposition, characteristic of all Zn-Fe group metal alloys, explained by Brenner [6]. Hence, Zn-Ni alloy coating at 3.0 A dm^{-2}, represented as (Zn-Ni)$_{3.0/mono}$ has been taken as the optimal coating configuration for monolayer deposition from the proposed bath.

Table 2. *Corrosion data for monolayer Zn-Ni alloy coatings developed at different c.d.'s.*

j A dm^{-2}	Content of Ni Wt. %	$-E_0$ V *vs.* SCE	i_{corr} μA cm^{-2}	CR × 10^{-2} mm y^{-1}
1.0	2.62	1.019	17.62	23.42
2.0	4.05	1.086	13.87	18.44
3.0	7.95	1.105	10.88	14.46
4.0	8.07	1.162	12.53	16.66

Optimization of cyclic cathode current densities (CCCD's) and number of layers in CMMA coatings

In the present work improving the corrosion resistance of monolayer Zn-Ni alloy by multilayer technique is guided by the following principles [23-25]:

I. Periodic change in current density (c.d.) allows the growth of coatings having periodic change in its chemical composition.

II. The corrosion resistance property of multilayer coatings or any functional property in general reaches its maximum value when thickness of the individual layers reaches optimal nanoscale.

The amplitude of compositional modulation diminishes rapidly when layer thicknesses below certain limit (about 50 nm).

Nanoscale multilayer coatings with alternate layers of alloys of different composition generally show unique properties than bulk materials when each individual layer thickness is below ~100 nm. Different layered combinations offer unique mechanical, optical, magnetic and electronic properties including improved good corrosion resistance. In all these cases, the properties of the multilayer coatings depend on two factors, namely the *composition* and *thickness* of the individual layers [26-28]. Hence by proper setting up of composition (by selection of cyclic current density, CCCD's) and thickness (by fixing the time for each layer deposition) of the individual layer it is possible to optimize the coating configuration [29,30]. Hence multilayer coatings have been accomplished under different combination of CCCD's and individual layer thickness. The experimental procedure for optimization of coating configuration is explained below.

Multilayer coatings having 10 layers (arbitrarily chosen) have been developed at different CCCD's, namely at 1.0/3.0 and 2.0/4.0 A dm^{-2}, using same binary alloy bath. The coatings at different sets of CCCD's were developed in order to try different possible modulations in composition of individual layers, and their corrosion behaviours were studied, and corrosion data are reported in Table 3.

Table 3. *Corrosion rates of CMMA Zn-Ni coatings having 10 layers at different CCCD's in comparison with (Zn-Ni)$_{3.0/mono}$ developed from same bath.*

CCCD's A dm^{-2}	$-E_0$ V *vs.* SCE	$i_{corr.}$ μA cm^{-2}	CR × 10^{-2} mm y^{-1}
(Zn-Ni)$_{1.0/3.0/10}$	1.085	8.416	11.18
(Zn-Ni)$_{2.0/4.0/10}$	1.108	6.606	8.78
(Zn-Ni)$_{3.0/mono}$	1.105	10.88	14.46

It may be noted that both (Zn-Ni)$_{1.0/3.0/10}$ and (Zn-Ni)$_{2.0/4.0/10}$ coatings exhibited less CR compared to (Zn-Ni)$_{3.0/mono}$ coating. Hence, by choosing the above CCCD's, multilayer coatings with higher degree of layering, *i.e.* with 30, 60, 120, 300 and 600 layers have been developed by proper setting up of the power source. It may be noted the CR's decreased drastically with an increase in number of layers in both sets of CCCD's. It further indicates that improved corrosion property is not the unique property of the layer composition; instead, the combined effect of composition modulation and thickness of individual layers, or the number of layers. It should be noted that CR decreased only up to 300 layers and then started increasing, *i.e.* 600 layers as shown in Table 4.

Table 4. *Effect of layering on corrosion behavior of CMMA (Zn-Ni)$_{1.0/3.0}$ and (Zn-Ni)$_{2.0/4.0}$ coating in comparison with monolayer (Zn-Ni)$_{3.0/mono}$ deposited from same bath at 303K*

Coating configuration	Number of layers	Average thickness of each layer, nm	E_0 V *vs.* SCE	i_{corr} μA cm^{-2}	CR × 10^{-2} mm y^{-1}
	10	1500	1.085	8.416	11.18
	30	750	1.037	7.125	9.47
(Zn-Ni)$_{1.0/3.0}$	60	250	1.056	4.531	6.02
	120	125	1.067	2.357	3.13
	300	50	1.041	1.319	1.75
	600	25	1.050	3.577	4.75
	10	1500	1.108	6.606	8.78
	30	750	1.085	4.989	6.63
(Zn-Ni)$_{2.0/4.0}$	60	250	1.050	2.125	2.82
	120	125	1.007	1.056	1.40
	300	50	1.135	0.31	0.41
	600	25	1.036	1.975	2.61
(Zn-Ni)$_{3.0/mono}$	Monolayer	15000	1.128	10.78	14.33

Further, though decrease of CR was found in both sets of CCCD's, the CR corresponding to (Zn-Ni)$_{2.0/4.0}$ at 300 layers is the least, it has been taken as the optimal configuration for peak performance against corrosion, and is represented as $(Zn-Ni)_{2.0/4.0/300}$. The CR found to increase at higher degree of layering, *i.e.* at 600 layers is due to shorter relaxation time for redistribution of metal ions at the diffusion layer, during plating. It should be noted that under optimal condition, the total thickness of CMMA $(Zn-Ni)_{2/4/300}$ coating was found to be ~15 μm, calculated from thickness Tester (Coatmeasure Model M&C), verified by Faraday law. Then from the total thickness and number of layers allowed to form (300), it is predicted that the average thickness of each layer is 50 nm.

Thermal stability of the monolayer and multilayer coatings

Thickness and Hardness of the coatings

The thickness and hardness of coatings were found to be decreased on heat treatment in both $(Zn-Ni)_{3.0/mono}$ and $(Zn-Ni)_{2.0/4.0/300}$ coatings. In the case of $(Zn-Ni)_{2.0/4.0/300}$ coatings the decrease may be attributed to the structural changes, evidenced by XRD study. The decrease is more pronounced in case of monolayer when compared to the multilayer Zn-Ni alloy coatings. The thickness of the coatings is directly related to the high tensile residual internal stresses, which result from the presence of Ni in the alloy [31]. The drop in the coating thickness with the temperature may be attributed to the iron enrichment caused by the formation of intermetallic Zn/Fe compounds due to the inter-diffusion at the coating/substrate interface. It was predicted that the iron enrichment in the interfacial region pushes nickel toward the surface [32]. The decrease of thickness and hardness with increase in temperature observed in case of $(Zn-Ni)_{2.0/4.0/300}$ is shown, in comparison with that of $(Zn-Ni)_{3.0/mono}$ coating in Table 5.

Table 5. *Effect of temperature on thickness, hardness and corrosion behavior of CMMA $(Zn-Ni)_{2.0/4.0/300}$ coatings, in comparison with monolayer $(Zn-Ni)_{3.0/mono}$ deposited from same bath.*

CCCD's A dm^{-2}	Treated temperature, °C	Thickness, μm	Vicker hardness, V_{100}	$-E_{corr}$ V *vs.* SCE	i_{corr} μA cm^{-2}	CR×10^{-2} mm y^{-1}
	30	18.9	212	1.003	0.31	0.41
	100	16.7	189	1.022	2.14	2.84
$(Zn-Ni)_{2.0/4.0/300}$	200	14.3	183	1.024	2.18	2.89
	300	9.1	154	0.995	3.73	4.95
	400	5.4	139	0.865	4.52	6.00
	30	17.8	202	1.128	10.78	14.33
$(Zn-Ni)_{3.0/mono}$	200	9.3	145	1.053	21.68	28.82
	400	4.8	119	1.070	26.53	35.27

Further, in the case of monolayer Zn-Ni coating the decrease of hardness may be attributed to the fact that the dislocation sources are active under the stress field (applied in the form of heat) [33]. The decrease in the residual stress is also reasoned to be responsible for decrease in hardness of the alloy coating with thermal treatment [34]. The CMMA coatings provide stress free environment, because the attractive forces tend to bridge the gap between successive layers. However, with heat treatment the interaction with the substrate increased, and hence stress developed.

Corrosion study

Cyclic polarization study

The corrosion resistance exhibited by Zn-Ni coatings (both monolayer and multilayer) can be better understood by Tafel polarization method as shown in Figure 2a, and corresponding corrosion parameters are reported in Table 5. Cyclic polarization study over a potential range of -1.25V to -0.5V was studied and is shown in Figure 2b. The anodic current was made to move from negative to positive, and then reversed. Cyclic polarization curves shows that there is no much significance of the corrosion product formation with regard to the corrosion protection of the alloy. The corrosion current value was found to be always higher than that of forward scanning, indicating the dissolution of oxide film had occurred in the forward scanning and self-repairing occurred during the process of backward scanning. CMMA Zn-Ni coating with optimal configuration, $(Zn-Ni)_{2.0/4.0/300}$ showing the least CR was opted for examining the thermal stability of the coatings. The CR of the Zn-Ni CMMA coating in 5% NaCl solution, before and after heat treatment of 100 °C represented as $(Zn-Ni)_{2.0/4.0/300}$ and $(Zn-Ni)_{2.0/4.0/300/100 °C}$), was found to be, respectively about 35 times and 5 times, respectively, more corrosion resistive than conventional $(Zn-Ni)_{3.0/mono}$ alloy coating as shown in Figure 2b.

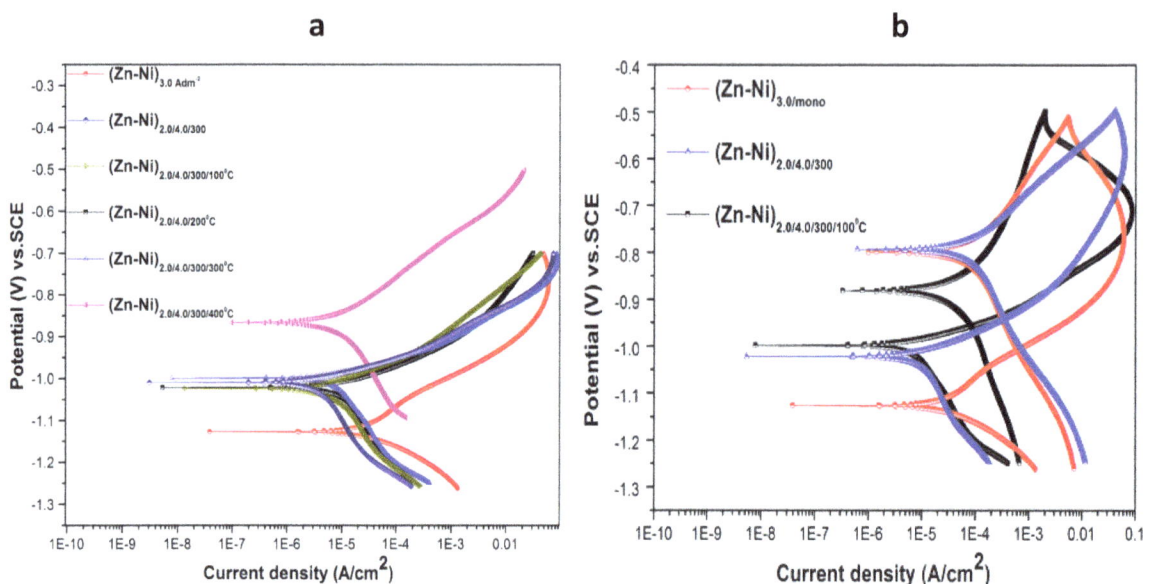

Figure 2. *Polarization behavior of CMMA $(Zn-Ni)_{2.0/4.0/300}$ coatings: a) after treatment at different temperature b), Cyclic polarization behavior of $(Zn-Ni)_{3.0/mono}$, CMMA $(Zn-Ni)_{2.0/4.0/300}$ and CMMA $(Zn-Ni)_{2.0/4.0/300/100°C}$.*

Thus the corrosion resistance of multilayered Zn-Ni alloy coating has also decreased with heat treatment. However, the decrease in CR's were more significant in the case of multilayer Zn-Ni alloy coating compared to monolayer coating after heat treatment. It is due to the fact that the electroplated samples attained intrinsically different surfaces in terms of their electrochemical properties. However, increase in the CR's is not in proportion of the structural changes observed, as will be discussed with XRD analysis.

Electrochemical impedance spectroscopy study

The electrode process involved in double layer capacitance and corrosion behaviors can be better understood by electrochemical impedance spectroscopy (EIS) method. The small amplitude signals from the test specimen are considered in EIS method. The Nyquist plot is the type of the plot in which the data is plotted as imaginary impedance, Z_{img} *vs.* real impedance Z_{real} with the provision to distinguish the contribution of polarization resistance (R_p) verses solution resistance (R_s) [35]. EIS signals of Zn-Ni monolayer coating (at optimal c.d, *i.e.* 3.0 A dm^{-2}) compared with the $(Zn-Ni)_{2.0/4.0/300}$ coating before and after heat treatment (at different temperatures) is shown on

Figure 3. Progressive decrease of polarization resistance of CMMA coatings with annealing temperature supports the reduced corrosion resistance, may be due to diffusion of layers. However, corrosion rate of monolayer coatings are more than multilayer coatings under all degree of layering.

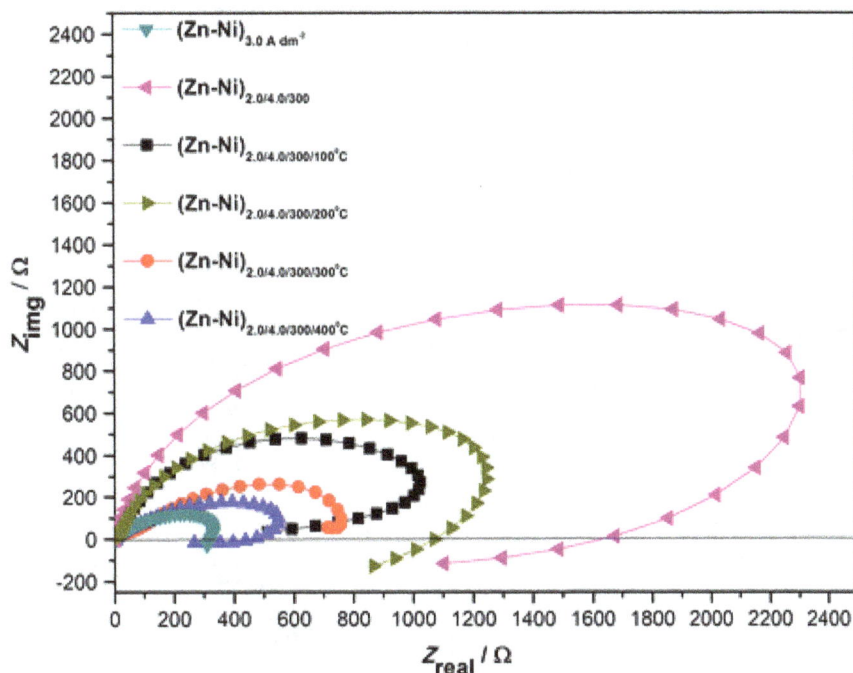

Figure 3. *Electrochemical impedance response (Real vs. Imaginary reactance values) displayed by CMMA (Zn-Ni)$_{2.0/4.0/300}$ after heat treatment at 4 different set temperatures.*

It may be noted that all the coatings exhibits one capacitive loop. However, at low frequency limit the capacitive behaviour of the double layer tends to exhibit the inductive character, indicated by decrease of both Z_{real} and Z_{img}. The gradual decrease in the diameter of semicircle indicates that the polarization resistance, R_P decreases with increase in temperature as shown in Figure 3. The (Zn-Ni)$_{2.0/4.0/300}$ coating showed maximum impedance, due to accumulation of corrosion products at the electrode surface acting as barrier. The negative value of imaginary impedance, Z_{img} at lower frequencies observed in coatings treated for higher temperatures are attributed to the inductance behaviour caused by the change in corrosion potential at the interface [36,37]. However, the radii of both capacitive loop and inductive character decreased progressively with raise in temperature.

Surface morphology

Figure 4 displays the SEM image for surface morphology of (Zn-Ni)$_{2.0/4.0/300}$ alloy coatings after heat treatment at different temperatures. It may be noted that the surface non-homogeneity increased with an increase of treatment temperature. However, Zn-Ni alloy coatings are found to adhere onto the substrate as fine particles with compact arrangement. The increased compactness with the heat treatment was also predicted to be main reason for good appearance and high hardness [38].

The cross sectional view of CMMA (Zn-Ni)$_{2.0/4.0/10}$ coating before and after heat treatment (100 °C) is shown in Figure 5. It may be noted that distinctly visible alloy layers (Figure 5a) become fused (Figure 5b) due to diffusion of layers upon heat treatment. Thus thermal treatment of electrodeposited multilayer Zn-Ni alloy coatings leads to both structural and behavioral changes.

Figure 4. *SEM images of CMMA (Zn-Ni)$_{2.0/4.0/300}$ coatings after treatment at different temperatures: (a) 100°C (b) 200°C (c) 300°C and (d) 400°C.*

Figure 5. *SEM images across the cross section of CMMA (Zn-Ni)$_{2.0/4.0/10}$ coatings before heat treatment (a), and after heat treatment at 200°C (b).*

XRD study

The XRD patterns of (Zn-Ni)$_{3.0/mono}$ and (Zn-Ni)$_{2.0/4.0/300}$ alloy coating after heat treatment, deposited from same bath is given in Figure 6 and 7 respectively. Variation in the surface morphology of the monolayer coatings after heat treatment is supported by the variation in XRD peaks. It has been reported that the phases obtained by the Zn-Ni coatings up to 13 % nickel do not correspond to that reported on the thermodynamic phase diagram [39]. It may be noted that the reflection corresponding to Zn(101), γ-phase (Ni$_5$Zn$_{21}$), Zn(103) phases and Ni$_3$Zn$_{22}$(335) as well was observed in monolayer coating deposited at optimal c.d. *i.e.*, 3.0 A dm^{-2}. However, upon heat treatment there was hardly any difference in the phases observed in the ((Zn-Ni)$_{3.0/mono}$ alloy coatings before and after heat treatment up to 200 °C. However, at temperatures higher than 200 °C additional phases were observed corresponding to the intermetallic Zn/Fe compound (Figure 6). The formation of this phase occurs due to inter diffusion in the interface region between the Zn-Ni alloy and the steel substrate. ZnO phase formation at temperatures higher than 200 °C was attributed to the metal contact with the ambient oxygen. The X-ray diffraction line broadening at temperatures higher than 200 °C may be related to the increase in the corrosion current value, which is reported to be caused by the lattice strains [40].

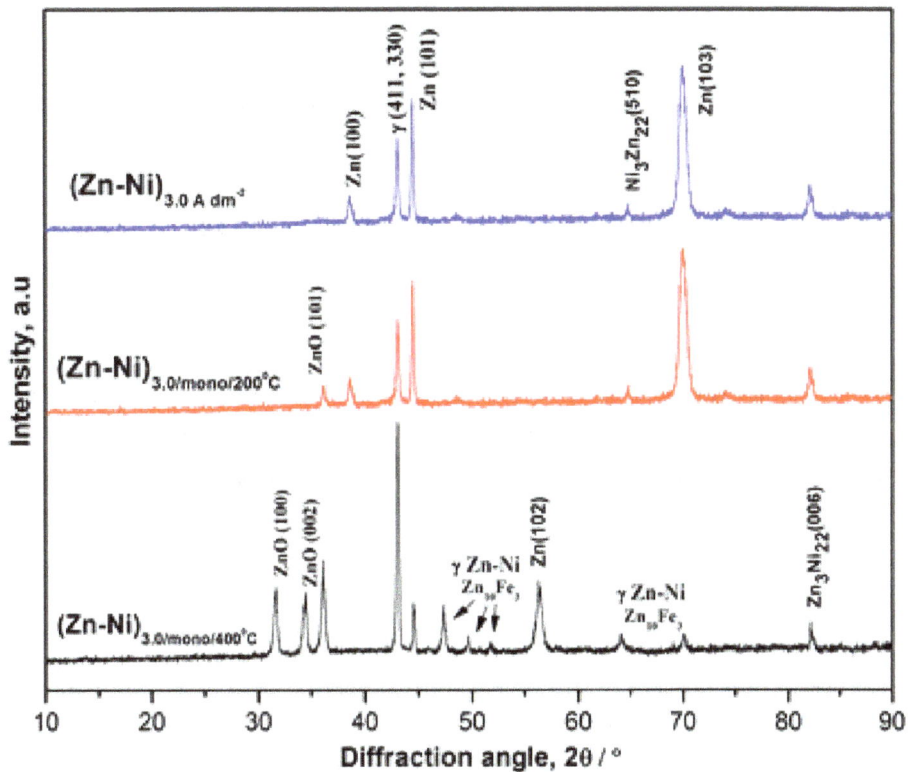

Figure 6. *XRD patterns of the (Zn-Ni)3.0/mono coatings after heat treatment at different temperatures (200°C and 400°C) deposited from the same bath.*

The reflection corresponding to Zn(101), γ-phase (Ni$_5$Zn$_{21}$) and Ni$_3$Zn$_{22}$(510) was highly suppressed in the case of (Zn-Ni)$_{2.0/4.0/300}$ coating on heat treatment at 4 different set temperatures. However at temperatures higher than 300 °C, the coating has shown weak signal corresponding to Zn(100), Zn(102) and γ Zn-Ni phases, although Zn(103) phase completely disappeared. The least CR exhibited by (Zn-Ni)$_{2.0/4.0/300/100}$ was attributed to the ratio of the phase structure corresponding to Zn(103) and weak signals corresponding to γ-phase of Zn (411, 330), Zn(101) and Zn$_3$Ni$_{22}$(006) phase. The comparison between the XRD signals of the Zn-Ni monolayer and CMMA coating on post-heat treatment reveals the fact that the exhibition of lesser CR of the Zn-Ni CMMA coating upon heat treatment to that of the Zn-Ni monolayer coating deposited at optimal c.d. *i.e.*, 3.0 A dm^{-2}, was attributed to the diffusion of the layered structure and formation of different phase structure ratio as shown on Figure 6 and Figure 7.

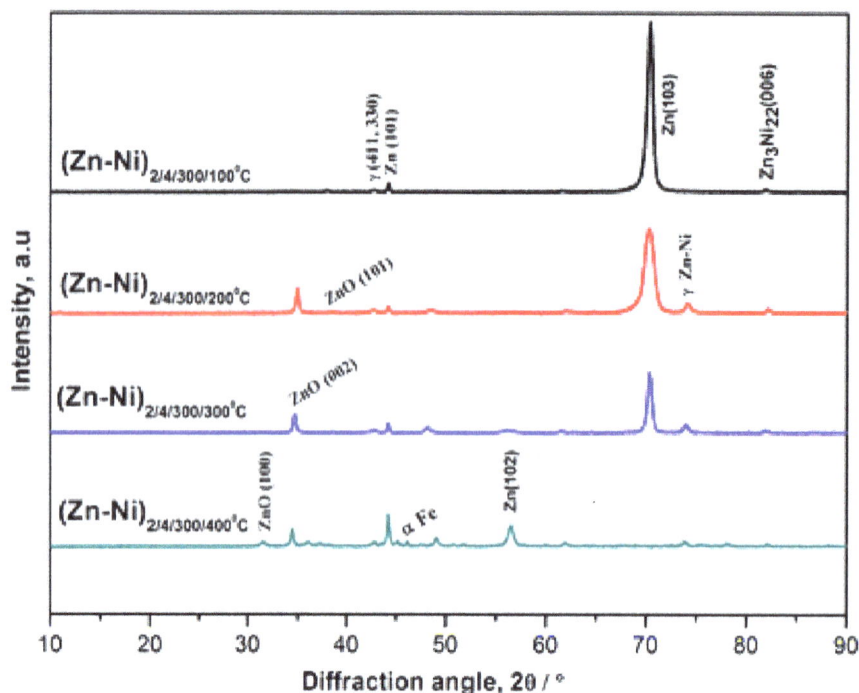

Figure 7. *XRD patterns of the CMMA (Zn-Ni)2.0/4.0/300 coatings after heat treatment at different temperature (100°C, 200°C, 300°C and 400°C), deposited from the same bath.*

Conclusions

Based on the experimental investigation on development and characterization of CMMA Zn-Ni alloy coatings on mild steel following observations were made as conclusions:

1. The coating configuration in terms CCCD's and number of layers have been optimized for deposition of the most corrosion resistant coatings from acid chloride bath using glycerol and gelatin as additives.

2. The decrease in thickness and hardness of both monolayer and multilayer coatings due to heat treatment were due to the active dislocation sources and decrease in the residual stress.

3. Progressive decrease of polarization resistance of CMMA coatings with annealing temperature supports the reduced corrosion resistance, may be due to diffusion of layers. However, corrosion rate of monolayer coatings are more than multilayer coatings under all degree of layering.

4. The corrosion resistance of multilayer coatings increased only up to certain number of layers and then decreased due to interlayer diffusion.

5. Thermal treatment of electrodeposited (both monolayer and multilayer) Zn-Ni alloy coatings led to significant structural change of the alloy, supported by SEM and XRD study.

6. A small increase of corrosion rates of Zn-Ni alloy due to annealing is related to the formation of a more ordered lattice of the alloy.

7. However, the increase of corrosion rates, due to annealing is not in proportion of the structural changes of the alloy occurred.

8. The corrosion resistance of both multilayered and monolayer coatings decreases with heat treatment. However, it is more pronounced in case of monolayer coatings. It is due to the fact that the electroplated samples attained intrinsically different interfacial structures.

9. The structural changes due to heat treatment of monolayer Zn-Ni alloy coatings continue to exist even in multilayer coatings. Further, the structural changes in alloy composition (both monolayer and multilayer) are found to be a function of annealing temperatures.

Acknowledgement: *Mr. Vaishaka R.Rao acknowledges National Institute of Technology Karnataka (NITK), Surathkal for financial support in the form of Institute Fellowship.*

References

[1] G. Injeti, B. Leo, *Adv. Mater.* **9** (2008) 1–11.

[2] P. Ganesan, S. P. Kumaraguru, B. N. Popov, *Surf. Coat. Technol.* **20** (2007) 7896– 7904.

[3] V. Thangaraj, N. Eliaz, A. C. Hegde, *J. Appl. Electrochem.* **39** (2009) 339–345.

[4] S. Yogesha, A. C. Hegde, *J. Mater. Process. Technol.* **211** (2011) 1409-1415.

[5] W. Blum, *Trans. Am. Electrochem Soc.* **40** (1921) 307-320.

[6] A. Brenner, *Electrodeposition of Alloy,* Academic Press., New York, p. 194.

[7] D. Tench, J. White, Tensile, *J. Electrochem Soc.* **138** (1991) 3757-58.

[8] Zhongda Wu, L. Fedrizzi, P.L. Bonora, *Surf. Coat. Technol.* **85** (1996) 170-174.

[9] R. Ramanauskas, I. Muleshkova, L. Maldonado, P. Dobrovolskis, *Corros. Sci.* **40** (1998) 401-410.

[10] M. Gavrila, J.P. Millet, H. Mazille, D. Marchandise, J.M. Cuntz, *Surf. Coat. Technol.* **123** (2000) 164–172.

[11] J.B. Bajat, Z. Kacarevic-Popovic, V.B. Miskovic-Stankovic, M.D. Maksimovic, *Prog. Org. Coat.* **39** (2000) 127–135.

[12] R. Fratesi, G. Roventi, *Surf. Coat. Technol.* **82** (1996) 158-164.

[13] Jing-yin Fei, G.D. Wilcox, *Surf. Coat. Technol.* **200** (2006) 3533 – 3539.

[14] M. Rahsepar, M.E. Bahrololoom, *Corros. Sci.* **51** (2009) 2537–2543.

[15] K. Venkatakrishna, A. C. Hegde, *J. Appl. Electrochem.* **40** (2010) 2051–2059.

[16] A. Maciej, G. Nawrat, W. Simka, J. Piotrowski, *Mater. Chem. Phys.* **132** (2012) 1095–1102.

[17] Y. Liao, D.R. Gabe, G.D. Wilcox, *Plat. Surf. Finish.* **85** (1998) 88-91.

[18] J.-L. Chen, J.-H. Liu, S.-M. Li, M. Yu, *Mater. Corros.* **63** (2012) 607-613.

[19] S. Yogesha, R.S. Bhat, K. Venkatakrishna, G. P. Pavithra, Y. Ullal, A.C. Hegde, *Synth. React. Inorg. Met.-Org. Chem.* **41** (2011) 65-71.

[20] R.S. Bhat, K.R. Udupa, A.C. Hegde, *Trans. Inst. Met. Finish.* **89** (2011) 268-274.

[21] V. Thangaraj, K. Ravishankar, A.C. Hegde, *Chin. J. Chem.* **26** (2008) 2285-2291.

[22] R. S. Bhat, K. U. Bhat, A. C. Hegde, *Prot. Met. Phys. Chem. Surf.* **47** (2011) 645-653.

[23] F. Jing-yin, L. Guo-zheng, X. Wen-li, W. Wei-kang, *J. Iron. Steel Res. Int.* **13** (2006) p. 61-67.

[24] R.S. Bhat, A.C. Hegde, *Surf. Eng. Appl. Electrochem.* **47** (2011) 112-119.

[25] B.N. Popova, S. Kumaragurua, P. Ganesana, *J. Electrochem Soc.* **1** (2006) 87-96.

[26] K.K. Chattopadhyay, A.N. Banerjee, Introduction to Nanoscience and Technology, Connaught Circus, New Delhi, India, 2009, p. 155.

[27] V.K. Varadan, A.S. Pillai, D. Mukherji, M. Dwivedi, L. Chen, Nanoscience and Nanotechnology in Engineering, World Scientific Publishing Co. Pvt. Ltd. Singapore, 2010, p. 170.

[28] D.R. Baer, P.E. Burrows, A.A. El-Azab, *Prog. Org. Coat.* **47** (2003) 342-356.

[29] K. Venkatakrishna, A. C. Hegde, *Mater. Manuf. Processes.* **26** (2011) 29–36.

[30] S. Yogesha, K. Venkatakrishna, A. C. Hegde, *Anti-Corros. Methods Mater.* **58** (2011) 84-89.

[31] H.J.C. Voorwald, I.M. Miguel, M.P. Peres, M.O.H. Cioffi, *J. Mater. Eng. Perform.* **14** (2005), 249-257.

[32] J. Kondratiuk, P. Kuhn, E. Labrenz, C. Bischoff, *Surf. Coat. Technol.* **205** (2011) 4141-4153.

[33] L. P. Bicelli, B. Bozzini, C. Mele, L. D'Urzo, *Int. J. Electrochem. Sci.* **3** (2008) 356 – 408.

[34] B.K. Prasad, *Mater. Sci. Eng.* **A277** (2000) 95-101.

[35]   V. M. Huang, Shao-Ling Wu, M. E. Orazem, N. Pebere, B. Tribollet, V. Vivier, *Electrochim. Acta.* **56** (2011) 8048-8057.

[36]   U.C. Nwaogu, C. Blawert, N. Scharnagl, W. Dietzel, K.U. Kainer, *Corros. Sci.* **52** (2010) 2143–2154.

[37]   R. Ghosh, D.D.N. Singh, *Surf. Coat. Technol.* **201** (2007) 7346– 7359.

[38]   M. Pushpavanam, S.R. Natarajan, K. Balakrishan, L.R. Sharma, *J. Appl. Electrochem.* **21** (1991) 642-645.

[39]   C. Bories, J.P. Bonino, A. Rousset, *J. Appl. Electrochem.* **29** (1999) 1045-1051.

[40]   R. Ramanauskas, R. Juskenas, A. Kalinicenko, *J. Solid State Electrochem.* **8** (2004) 416-421.

9

Influence of operating parameters on electrocoagulation of C.I. disperse yellow 3

Djamel Ghernaout*,**,✉, Abdulaziz Ibraheem Al-Ghonamy**, Mohamed Wahib Naceur*, Noureddine Ait Messaoudene** and Mohamed Aichouni**

*Department of Chemical Engineering, University of Blida, PO Box 270, Blida 09000, Algeria

**Binladin Research Chair on Quality and Productivity Improvement in the Construction Industry; College of Engineering, University of Hail, PO Box 2440, Ha'il 81441, Saudi Arabia

✉Corresponding Author

Abstract

This work deals with the electrocoagulation (EC) process for an organic dye removal. The chosen organic dye is C.I. disperse yellow 3 (DY) which is used in textile industry. Experiments were performed in batch mode using Al electrodes and for comparison purposes Fe electrodes. The experimental set-up was composed of 1 L beaker, two identical electrodes which are separated 2 cm from each other. The main operating parameters influencing EC process were examined such as pH, supporting electrolyte concentration C_{NaCl}, current density i, and DY concentration. High performance EC process was shown during 45 min for 200 mg/L dye concentration at i = 350 A m^{-2} (applied voltage 12 V) and C_{NaCl} = 1 g L^{-1} reaching 98 % for pHs 3 and 10 and 99 % for pH 6. After 10 min, DY was also efficiently removed (86 %) showing that EC process may be conveniently applied for textile industry wastewater treatment. EC using Fe electrodes exhibited slightly lower performance comparing EC using Al electrodes.

Keywords:
Dye removal; aluminium; iron; current density; mechanism.

Introduction

The main problem of access to safe drinking water is continuous pollution of water resources by agriculture, urban waste and industry. In countries where water resources are relatively limited, treated wastewater reuse in agriculture has become an urgent necessity. The textile industry consumes considerable amounts of water in the dyeing and finishing. Effluents containing dyes can be toxic to the environment [1-4]. In addition, their presence in aquatic systems, even at low concentrations, is very visible. It reduces the penetration of light and has a detrimental effect on photosynthesis [5-7].

Therefore, the remediation of water contaminated by these chemicals is necessary both to protect the environment and for future reuse [8-12]. Therefore, several biological, physical and chemical methods are used for the treatment of industrial effluents with different efficeincies [13-15]. Electrochemical technologies, such as electrocoagulation technique (EC), seem to be well adapted to the textile industry wastewaters treatment [16-22].

This work is devoted to the study of the EC process for bleaching synthetic water containing an azo dye, C.I. disperse yellow 3 (DY), used in the Algerian textile industry and the assessment of its performance versus certain operating parameters.

Experimental

Experimental set-up

The EC tests were performed using an experimental set-up shown in Figure 1. In a 1000 mL beaker, filled with 500 mL synthetic dye solution (distilled water + DY + NaCl), two Al (or Fe in some experiments) 4×20 cm electrodes were immersed (active surface $S = 4 \times 10.5 = 42$ cm^2). The anode is connected to the positive pole and the cathode to the negative pole of the direct current power supply. The interelectrode distance is fixed at 2 cm. When the electric current is applied, the magnetic stirrer is started at an average velocity agitation.

Figure 1. *Photo of the EC experimental set-up.*

Electrodes cleaning

Before experiments, the Fe electrodes were prepared to avoid the presence of any impurity as follows: (1) polishing with abrasive paper; (2) rinsing with distilled water; (3) degreasing by means of a solution composed of: NaOH (25 g), Na$_2$CO$_3$ (25 g), K$_2$CO$_3$ (25 g) and distilled water (q.s.p. 1,000 mL); (4) rinsing with distilled water; (5) pickling in a solution of sulphuric acid H$_2$SO$_4$ at 20 % for 20 min at room temperature; and again (6) rinsing with distilled water. For Al electrodes: (1) rinse with distilled water and polish using abrasive paper, (2) clean in hydrochloric acid solution (HCl at 20 %) during 10 min, and (4) rinse with distilled water.

Prepared solutions

To prepare a solution of 200 mg L^{-1} dye, 0.2 g of the latter was poured into a 1 L flask and distilled water was added during stirring for better solubilisation. The initial pH was varied using a solution of 0.1 M HCl (acidic conditions) or NaOH (alkaline medium). The solution conductivity was increased by sodium chloride addition. All chemicals used were of analytical grade.

Methods

Once the EC test ends, the treated solutions were left to settle for 30 min in order to sediment the flocs formed. After decantation, and using a pipette, 25 mL of the solution were carefully collected for analysis.

The analyses done before and after treatment were as follows: pH, conductivity and ultraviolet (UV) absorbance (Shimadzu 1601, dual beam with 1 cm quartz vessel). The best UV absorbance long wave was found at 346 nm (UV_{346}). The DY removal was calculated using the relation (1):

$$R/\% = \frac{Ab_i - Ab_f}{Ab_f} \times 100 \tag{1}$$

where Ab_i and Ab_f were initial and final UV absorbances, respectively. All the tests were conducted at ambient temperature (20 °C).

Results and discussion

The aim of this work was to perform bleaching EC tests on dye synthetic solutions (distilled water+dye+NaCl) using EC process and evaluate its performance based on certain key parameters.

Influencing parameters on EC process

Common remarks

During EC tests, some common observations were:

- Aluminium dissolution according to Reaction (2):

$$Al \rightarrow Al^{3+} + e^- \tag{2}$$

- Production of H_2 gas bubbles at the cathode according to Reaction (3):

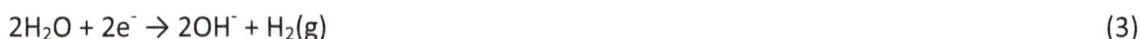

$$2H_2O + 2e^- \rightarrow 2OH^- + H_2(g) \tag{3}$$

- Production of O_2 gas bubbles at the anode according to Reaction (4):

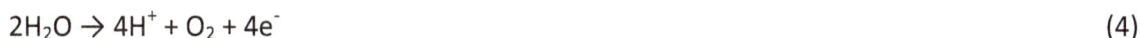

$$2H_2O \rightarrow 4H^+ + O_2 + 4e^- \tag{4}$$

- Flocs formation and their fixation on the $H_2(g)$ bubbles during their ascension to the solution surface as a white foam (Figure 2a and b). Indeed, anode dissolution generates coagulant species which destabilise the dye molecules forming flocs.

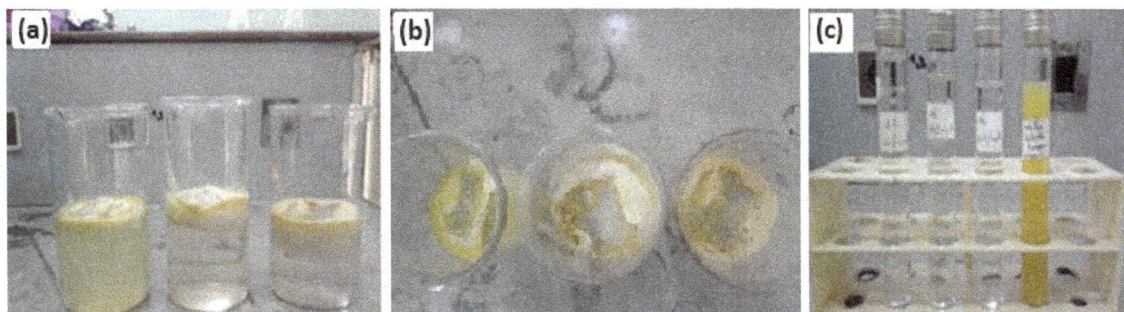

Figure 2. Foam formation: (a) face view, (b) top view. (c) Initial and final state of a 200 mg L^{-1} DY solution treated by EC during 1 h: from wright to left, initial solution, treated solution at 12, 8 and 4 V, respectively.

EC time

The EC efficiency is strongly influenced by the time residence in the electrochemical device. To study its effect, the EC period was varied from 5 to 75 min and the other parameters were kept constant. The results obtained are illustrated in Figure 3. The H_2 and O_2 release and flocs formation increased over time and the foam became thicker. From 30 min, the flocs settled and the solution became clearer.

As seen in Figure 3, the dye removal efficiency increased with electrolysis time until 45 min. After this time, EC performance decreased. Moreover, the good EC efficiencies were reached between 15 and 45 min.

The removal efficiency was directly dependent upon the metal concentration in solution [23-25]. The positive metallic species were produced by the Al anode neutralising the negative charges on the polluting molecules [26-30]. When the electrolysis duration was increased, the cationic species as well as metal hydroxide ($Al(OH)_{3(s)}$) concentrations increased [30-32]. Consequently, the pollutant removal increased [33-36].

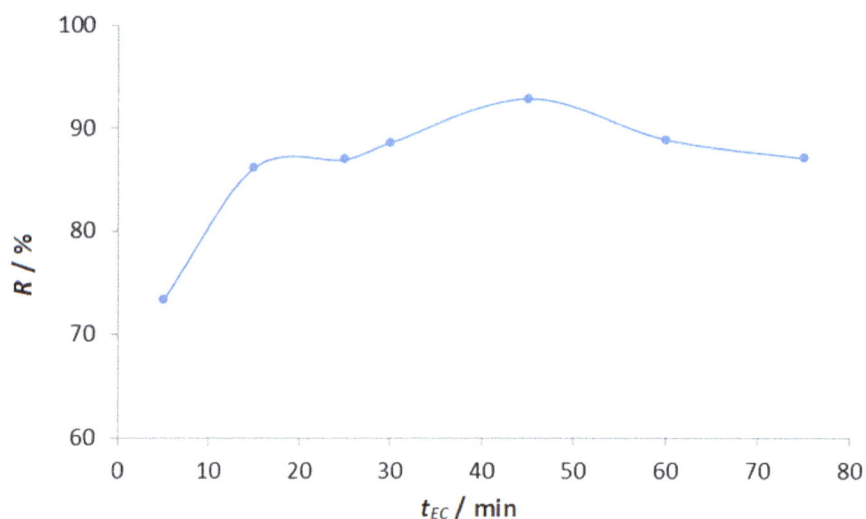

Figure 3. *Dye removal as a function of EC time (pH 6.5; $C_{NaCl} = 1$ g L^{-1}; $C_{DY} = 20$ mg L^{-1}; d = 2 cm).*

Electric current density

The electric current density is the most important parameter of the electrochemical process [37]. The electric current density effect on the dye removal was studied. The current intensities were 120, 250 and 350 mA corresponding to the applied voltages of 4, 8 and 12 V, respectively. The dye concentration was fixed at 20 and 200 mg/L. The other parameters were maintained constant (pH 6.5; $C_{NaCl} = 1$ g/L; $d = 2$ cm). The obtained results are shown in Figure 4.

For $I = 120$ mA ($i = 29$ A/m^2), the produced gas bubbles were small and the formed froth was thin. For $I = 250$ mA ($i = 60$ A/m^2), the gas emanation was medium and the formed froth became important. For $I = 350$ mA ($i = 83$ A/m^2), flocs settling became significant, the gas emanation became intense and the solution was transformed clear.

As seen in Figure 4, the electric current had a great effect on the dye removal especially for the first ten minutes. After 20 min, the electric current had a small effect. This is explained by the fact that the negative charge on the organic dye is neutralised after the Al^{3+} action on the dye molecules.

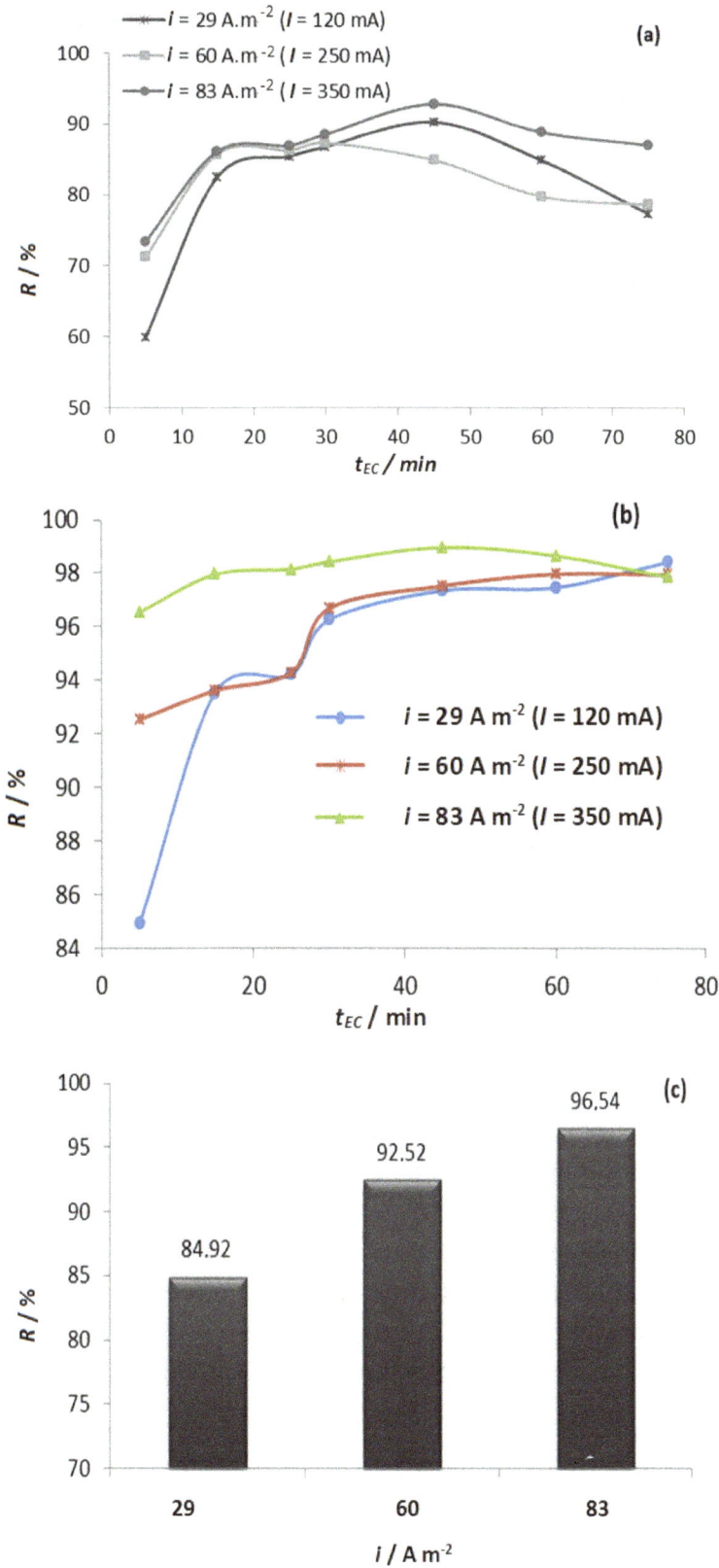

Figure 4. *Effect of the electric current i on the EC efficiency for DY removal (pH 6.5, d = 2 cm, C_{NaCl} = 1 g L^{-1})*
(a): C_{DY} = 20 mg L^{-1}; (b): C_{DY} = 200 mg L^{-1}; (c): C_{DY} = 200 mg.L^{-1}; t_{EC} = 5 min

Initial pH

The solution pH is an important factor influencing the EC performance [37]. This is due to the fact that pH determinates the metallic ions speciation, the chemical state of other species in the solution, and the formed products solubility. In order to examine the pH effect, the solution pH was adjusted to the values from 3 to 12 maintaining other parameters constant: U = 12 V (I = 83 A m^{-2}), C_{NaCl} = 1 g L^{-1}, d = 2 cm, t_{EC} = 30 min). The obtained results are shown in Figure 5.

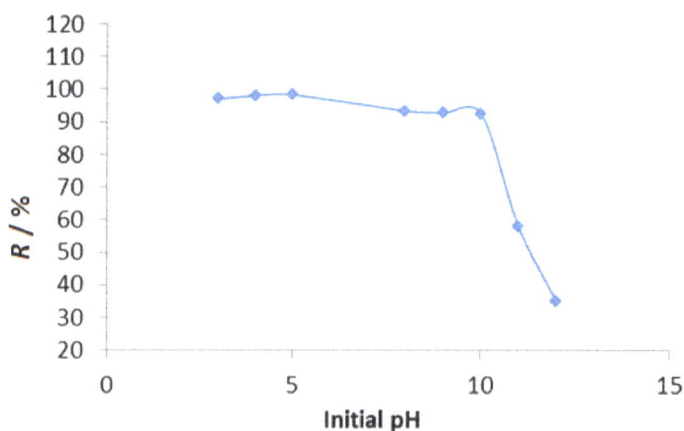

Figure 5. *Effect of the pH on the EC efficiency for DY removal (C_{DY} = 200 mg L^{-1}, U = 12 V (i = 83 A m^{-2}); t$_{EC}$ = 30 min; d = 2 cm).*

For the acidic medium, the solution colour became orange after the addition of HCl. From the cathode, there is an intense emanation of H$_2$ bubbles. From the anode, there is an important formation of O$_2$ bubbles. The foam and sediment formed are denser at the anode.

For the alkaline medium, there was formation of white sediment at the bottom of the beaker, and its volume increased with time in comparison with the acidic medium.

As seen in Figure 5, we noted that the dye removal was well performed between pH 3 and 10. Several researchers found that the best removal efficiency with aluminum electrodes was reached in the pH range between 3 and 9 [19,34-36].

In Figure 6, we chose four pH values: 3, 4, 6.5 and 10 with other parameters fixed in order to illustrate the pH effect.

We also followed the change in pH as a function of time; Figure 7 shows the obtained results.

The medium pH changed during the EC process. This change depended on the type of electrode material and the initial pH of the treated solution.

We note from Figure 7 an increase in pH in the case of solutions with pH < 7. This increase was probably due to the release of H$_2$ from the cathode and the formation of OH$^-$ according to the Reaction (3). Moreover, a decrease in pH in the case of solutions with pH > 7 was also noticed. This decrease was affected to the hydroxyl (OH$^-$) consumption according to Reaction (5):

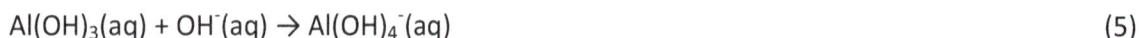

$$Al(OH)_3(aq) + OH^-(aq) \rightarrow Al(OH)_4^-(aq) \qquad (5)$$

Initial conductivity

Conductivity promotes the performance of electrochemical processes [26]. We chose NaCl as a supporting electrolyte. In order to determine its effect on the efficiency of bleaching of the synthetic solutions, we varied the concentration (0.25, 0.5, 1 and 1.5 g L^{-1}) while keeping the other parameters constant. The results are shown in Figure 8.

We have observed that (1) the gas production becomes higher with the increase in salt concentration, (2) the conductivity decreases during EC treatment and, (3) the formation of a small deposit on the anode.

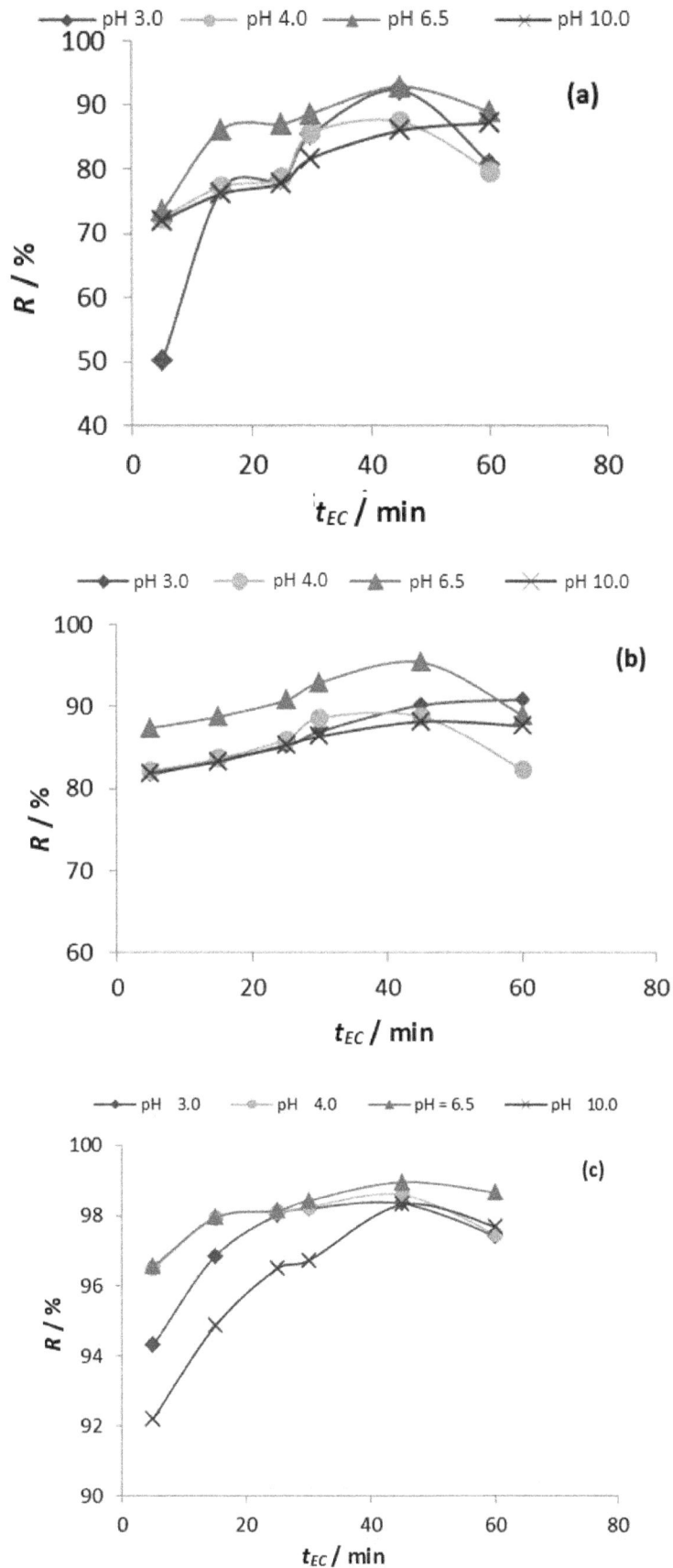

Figure 6. DY removal as a function of pH ($U = 12\ V$ ($i = 83\ A\ m^{-2}$), $t_{EC} = 30\ min$; $C_{NaCl} = 1\ g\ L^{-1}$; $d = 2\ cm$ (a) $C_{DY} = 20\ mg\ L^{-1}$; (b) $C_{DY} = 40\ mg\ L^{-1}$; (c) $C_{DY} = 200\ mg\ L^{-1}$

Figure 9 shows the solution conductivity as a function of time during EC process. We note that the conductivity decreased over time and the difference in the changes in pH was different due to the HCl and NaOH added during the pH adjustment before EC treatment.

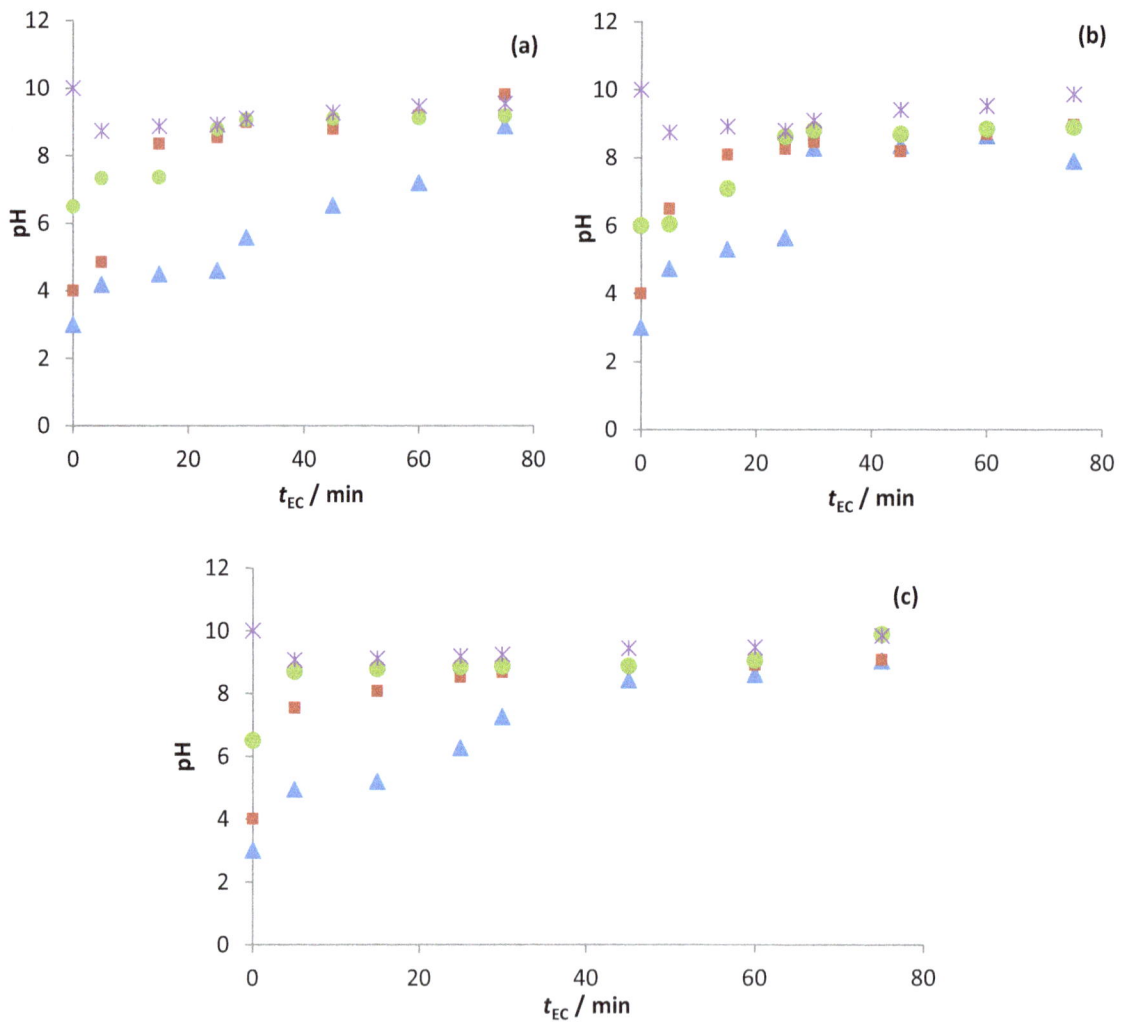

Figure 7. *Evolution of pH during EC treatment (same conditions as for Figure 6). (a) C_{DY} = 20 mg L^{-1}; (b) C_{DY} = 40 mg L^{-1}; (c) C_{DY} = 200 mgL^{-1}*

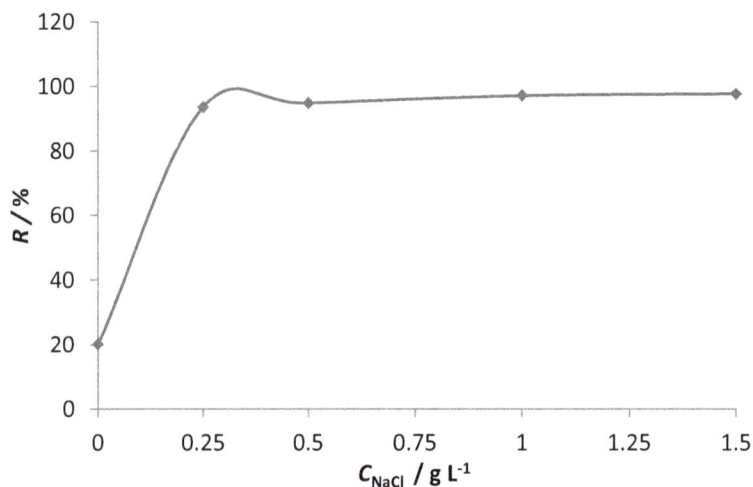

Figure 8. *Effect of the NaCl concentration on the EC efficiency U = 12 V (i = 83 A m^{-2}); t_{EC} = 30 min; d = 2 cm; C_{DY} = 200 mg L^{-1}*

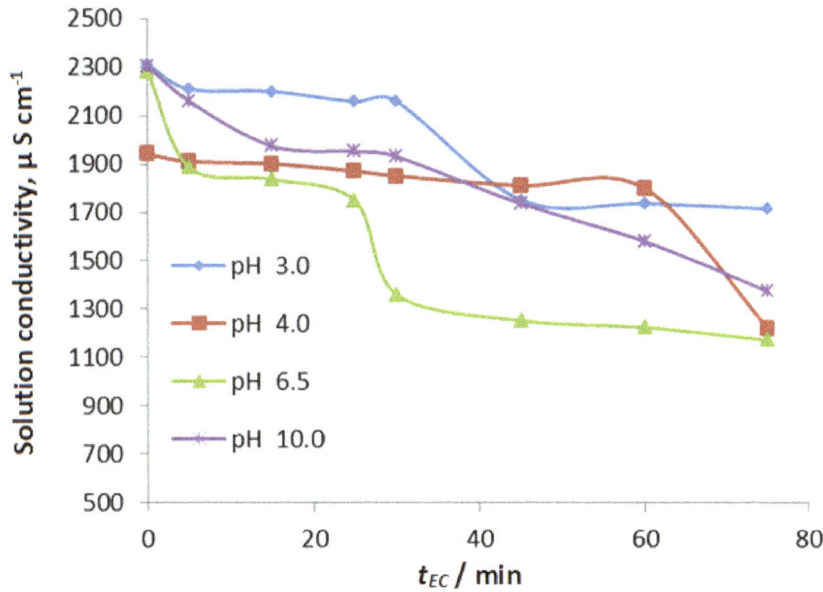

Figure 9. *Solution conductivity as a function of time during EC process*
(C_{NaCl} = 1 g L^{-1}; C_{DY} = 200 mg L^{-1}; d = 2 cm; U = 12 V (i = 83 A^{-2})

Inter-electrode distance

We varied the distance between the electrodes d = 0.8; 1; 1.5 and 2 cm while fixing the other factors. The results are shown in Figure 10. When increasing the inter-electrode distance, the EC efficiency also increased. This can be explained as follows: for *d* = 2 cm, there would be more probabilities to generate global flocs that are able to adsorb more dye molecules.

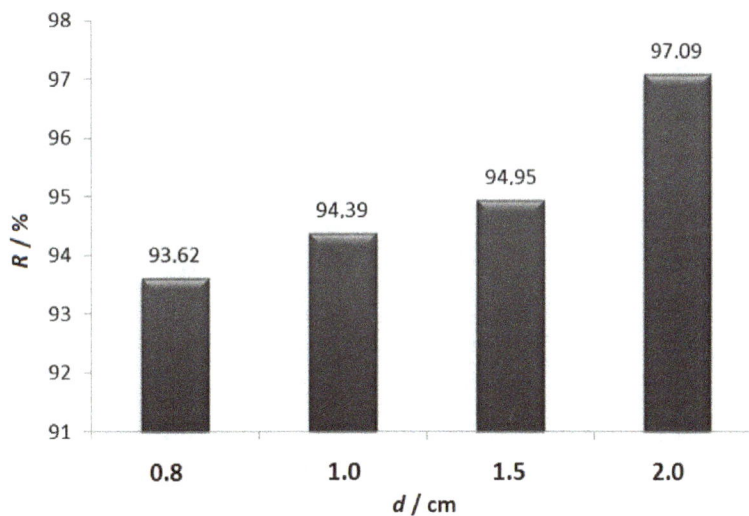

Figure 10. *Effect of the inter-electrode distance on the EC performance*
(C_{NaCl} = 1 g L^{-1}, C_{DY} = 200 mg L^{-1}, pH = 6.5, U = 12 V (i = 83 A m^{-2}).

DY concentration

The aim of this part is to determine whether the EC method was applicable to solutions with a range of concentrations from 20 to 500 mg L^{-1}. The solutions were electrolysed, for a treatment time of 45 min, at a fixed voltage U = 12 V (i = 83 A m^{-2}) and an inter-electrode distance of 2 cm. The results obtained are shown in Fig. 11.

Figure 11 shows that the EC method is effective in the range of selected concentrations. A yield of 97 % is reached at a dye concentration of 200 mg L^{-1}. The results obtained can be justified by the increased probability of contact with the dye molecules to aluminum hydroxide $Al(OH)_3$ to form flocs of large sizes; thereby facilitating their separation by their attachment to the bubbles of released gases at the electrodes.

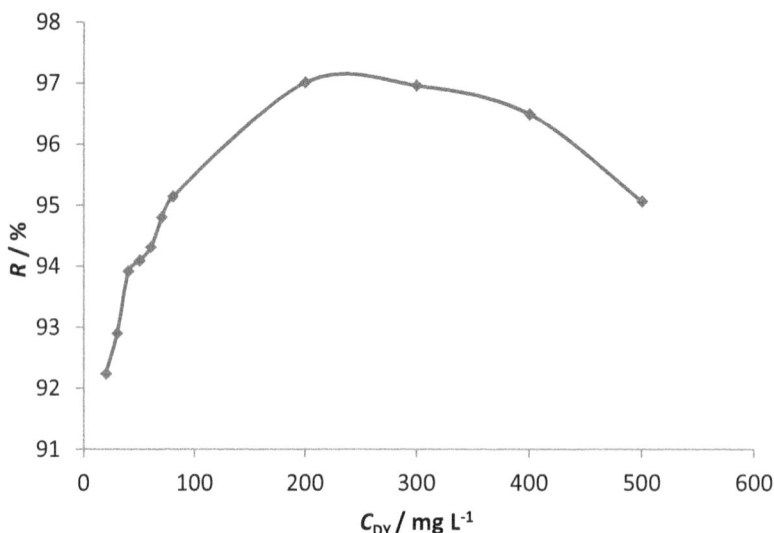

Figure 11. *EC performance as a function of DY concentration*
$C_{NaCl} = 1\ g\ L^{-1}$; *pH 6.5; d = 2 cm, U = 12 V (i = 83 Am^{-2})*

EC using Fe electrodes

To check if DY can be removed by EC using iron electrodes, some tests using the Al optimum conditions are performed. The results obtained are compared with those obtained with aluminum electrodes (Figure 12).

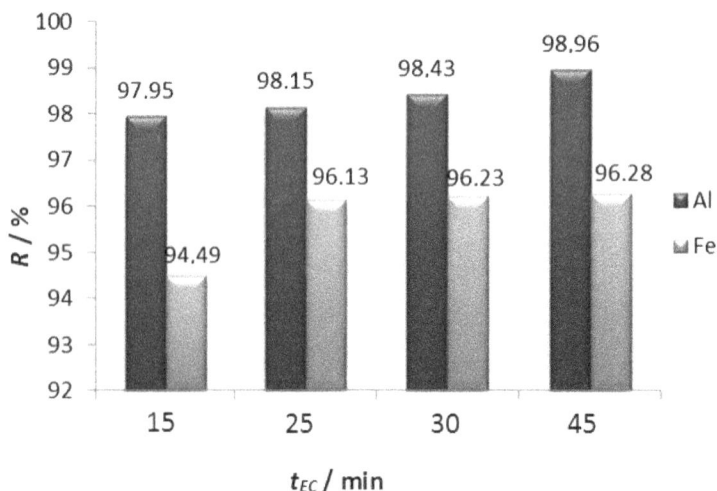

Figure 12. *EC using Al and Fe electrodes:* $C_{DY} = 200\ mg\ L^{-1}$; $C_{NaCl} = 1\ g\ L^{-1}$;
pH 6.5; d = 2 cm, U = 12 V (i = 83 A/m^2)

During EC treatment using Fe electrodes, (1) clouds of green flocs came from the anode surface and settled to the beaker bottom and, (2) on the solution surface, two layers of foam were observed: the first a red-brown color, and below, the second green.

We find that the rate of reduction of the dye increased with time until a yield of 96.28 % in the case of iron electrodes (Figure 12). Comparing the test results with aluminum electrodes (R = 98.96 %), we can say that iron is also effective in removing DY.

The reactions involved in the EC using iron electrodes are as follows: The iron, after oxidation in the electrolytic system, produces iron hydroxide $Fe(OH)_n(s)$, with n = 2 or 3; and two mechanisms have been proposed [6,12,14,15,24,38]:

- Mechanism 1 (green coloration, $Fe(OH)_{2(s)}$):

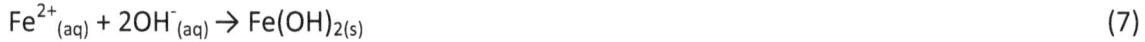

$$Fe_{(s)} \rightarrow Fe^{2+}_{(aq)} + 2e^- \tag{6}$$

$$Fe^{2+}_{(aq)} + 2OH^-_{(aq)} \rightarrow Fe(OH)_{2(s)} \tag{7}$$

- Mechanism 2 (brown coloration, $Fe(OH)_{3(s)}$):

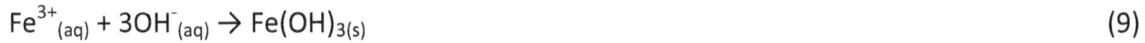

$$Fe^{2+}_{(aq)} \rightarrow Fe^{3+}_{(aq)} + 1e^- \tag{8}$$

$$Fe^{3+}_{(aq)} + 3OH^-_{(aq)} \rightarrow Fe(OH)_{3(s)} \tag{9}$$

The species $Fe(OH)_{n(s)}$ formed (by the two mechanisms) remained in the aqueous phase in the form of gelatinous suspension which can then remove the pollutants from the water (Figure 13), either by complexation, or by electrostatic attraction, followed by coagulation and flotation or sedimentation [24,25,27,28].

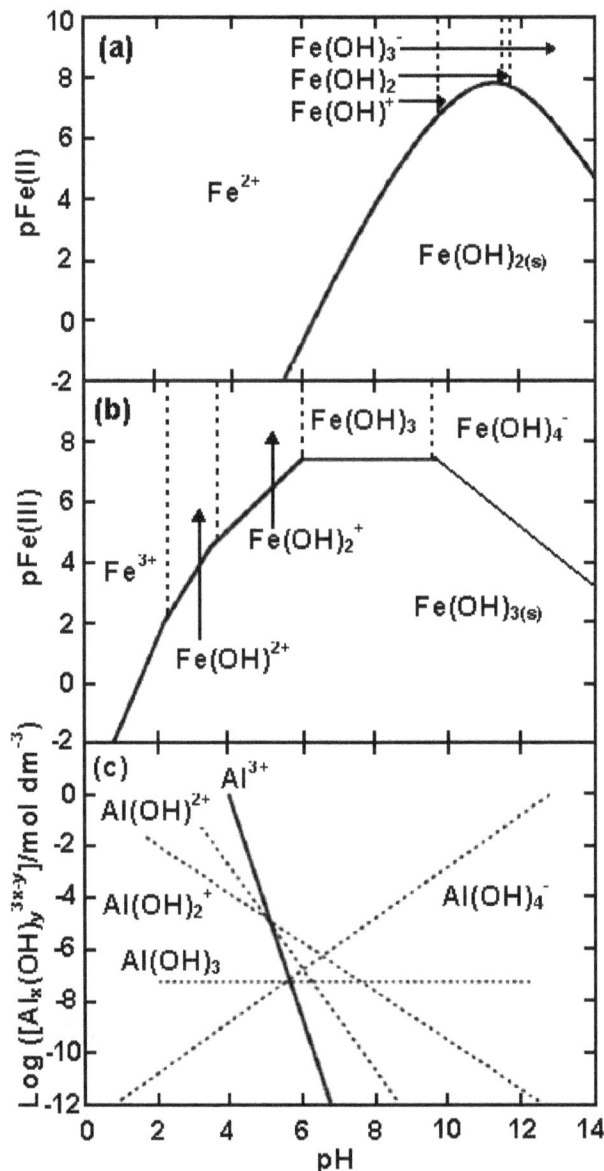

Figure 13. Predominance-zone diagrams for (a) Fe(II) and (b) Fe(III) chemical species in aqueous solution. The straight lines represent the solubility equilibrium for insoluble $Fe(OH)_2$ and $Fe(OH)_3$, respectively, and the dotted lines represent the predominance limits between soluble chemical species. (c) Diagram of solubility of Al(III) species as a function of pH [27,28].

Conclusions

Highly performant EC process is shown in dye removal during 45 min for 200 mg/L dye concentration at i = 350 A/m^2 (applied voltage 12 V) and C_{NaCl} = 1 g/L reaching 98 % for pH 3 and 10 and 99 % for pH 6. For 10 min, DY is also efficiently removed (86 %) showing that EC process may be well convenient for textile industry wastewater treatment. EC using Fe electrodes is slightly less performant than EC using Al electrodes.

Acknowledgements: *The present research work was undertaken by the Binladin Research Chair on Quality and Productivity Improvement in the Construction Industry funded by the Saudi Binladin Constructions Group; this is gratefully acknowledged. The opinions and conclusions presented in this paper are those of the authors and do not necessarily reflect the views of the sponsoring organisation.*

References

[1] A. R. Amani-Ghadim, S. Aber, A. Olad, H. Ashassi-Sorkhabi, *Chemical Engineering and Processing* **64** (2013) 68-78.

[2] W. Lemlikchi, S. Khaldi, M.O. Mecherri, H. Lounici, N. Drouiche, *Separation Science and Technology* **47** (2012) 1682-1688.

[3] M. S. Secula, B. Cagnon, T. F. de Oliveira, O. Chedeville, H. Fauduet, *Journal of the Taiwan Institute of Chemical Engineers* 43 (2012) 767-775.

[4] B. Merzouk, B. Gourich, K. Madani, Ch. Vial, A. Sekki, *Desalination* 272 (2011) 246-253.

[5] M. Kobya, E. Demirbas, O. T. Can, M. Bayramoglu, *Journal of Hazardous Materials* **B132** (2006) 183-188.

[6] A. Aleboyeh, N. Daneshvar, M. B. Kasiri, *Chemical Engineering and Processing* **47** (2008) 827-832.

[7] S. Zodi, B. Merzouk, O. Potier, F. Lapicque, J.-P. Leclerc, *Separation and Purification Technology* **108** (2013) 215-222.

[8] M. Eyvaz, M. Kirlaroglu, T. S. Aktas, E. Yuksel, *Chemical Engineering Journal* **153** (2009) 16-22.

[9] E. Pajootan, M. Arami, N. M. Mahmoodi, *Journal of the Taiwan Institute of Chemical Engineers* **43** (2012) 282-290.

[10] M.-C. Wei, K.-S. Wang, C.-L. Huang, C.-W. Chiang, T.-J. Chang, S.-S. Lee, S.-H. Chang, *Chemical Engineering Journal* **192** (2012) 37-44.

[11] C. Phalakornkule, P. Sukkasem, C. Mutchimsattha, *International Journal of Hydrogen Energy* 35 (2010) 10934-10943.

[12] M. Chafi, B. Gourich, A. H. Essadki, C. Vial, A. Fabregat, *Desalination* **281** (2011) 285-292.

[13] P. Cañizares, F. Martínez, M.A. Rodrigo, C. Jiménez, C. Sáez, J. Lobato, *Separation and Purification Technology* 60 (2008) 147-154.

[14] C. Gong, Z. Zhang, H. Li, D. Li, B. Wu,Y. Sun, Y. Cheng, *Journal of Hazardous Materials* 274 (2014) 465-472.

[15] J. Llanos, S. Cotillas, P. Cañizares, M.A. Rodrigo, *Water Research* **53** (2014) 329-338.

[16] S. Pulkka, M. Martikainen, A. Bhatnagar, M. Sillanpää, *Separation and Purification Technology* **132** (2014) 252-271.

[17] E-S. Z. El-Ashtoukhy, N.K. Amin, *Journal of Hazardous Materials* **179** (2010) 113-119.

[18] A. K. Golder, N. Hridaya, A. N. Samanta, S. Ray, *Journal of Hazardous Materials* **B127** (2005) 134-140.

[19] S. Aoudj, A. Khelifa, N. Drouiche, M. Hecini, H. Hamitouche, *Chemical Engineering and Processing* 49 (2010) 1176-1182.

[20] Y. Ş Yildiz, *Journal of Hazardous Materials* **153** (2008) 194-200.

[21]   C.-L. Yang, J. McGarrahan, *Journal of Hazardous Materials* **B127** (2005) 40-47.

[22]   J. G. Ibanez, M. M. Singh, Z. Szafran, *Journal of Chemical Education* **75** (1998) 1040-1041.

[23]   D. Ghernaout, S. Irki, A. Boucherit, *Desalination and Water Treatment* **52** (2014) 3256-3270.

[24]   D. Ghernaout, *Desalination and Water Treatment* **51** (2013) 7536-7554.

[25]   D. Ghernaout, M. W. Naceur, B. Ghernaout, *Desalination and Water Treatment* **28** (2011) 287-320.

[26]   D. Ghernaout, B. Ghernaout, *Desalination and Water Treatment* **27** (2011) 243-254.

[27]   D. Ghernaout, M.W. Naceur, A. Aouabed, *Desalination* **270** (2011) 9-22.

[28]   C. A. Martínez-Huitle, E. Brillas, *Applied Catalysis B: Environmental* 87 (2009) 105-145.

[29]   D. Ghernaout, A. Mariche, B. Ghernaout, A. Kellil, *Desalination and Water Treatment* **22** (2010) 311-329.

[30]   A. Saiba, S. Kourdali, B. Ghernaout, D. Ghernaout, *Desalination and Water Treatment* **16** (2010) 201-217.

[31]   D. Ghernaout, B. Ghernaout, *Desalination and Water Treatment* **16** (2010) 156-175.

[32]   D. Belhout, D. Ghernaout, S. Djezzar-Douakh, A. Kellil, *Desalination and Water Treatment* **16** (2010) 1-9.

[33]   D. Ghernaout, B. Ghernaout, A. Boucherit, M.W. Naceur, A. Khelifa, A. Kellil, *Desalination and Water Treatment* 8 (2009) 91-99.

[34]   D. Ghernaout, A. Badis, A. Kellil, B. Ghernaout, *Desalination* 219 (2008) 118-125.

[35]   D. Ghernaout, B. Ghernaout, A. Saiba, A. Boucherit, A. Kellil, *Desalination* **239** (2009) 295-308.

[36]   D. Ghernaout, B. Ghernaout, A. Boucherit, *Journal of Dispersion Science and Technology* **29** (2008) 1272-1275.

[37]   M. Muthukumar, M. Govindaraj, A. Muthusamy, G. Bhaskar Raju, *Separation Science and Technology* **46** (2011) 272-282.

[38]   C. Barrera-Díaz, B. Bilyeu, G. Roa, L. Bernal-Martinez, *Separation and Purification Reviews* **40** (2011) 1-24.

10

Phenolic compounds removal from mimosa tannin model water and olive mill wastewater by energy-efficient electrocoagulation process

Marijana Kraljić Roković⊠, Mario Čubrić and Ozren Wittine

Faculty of Chemical Engineering and Technology, University of Zagreb, Marulićev trg 19, Croatia

⊠Corresponding Author

Abstract

The objective of this work was to study the influence of NaCl concentration, time, and current density on the removal efficiency of phenolic compounds by electrocoagulation process, as well as to compare the specific energy consumption (SEC) of these processes under different experimental conditions. Electrocoagulation was carried out on two different samples of water: model water of mimosa tannin and olive mill wastewater (OMW). Low carbon steel electrodes were used in the experiments. The properties of the treated effluent were determined using UV/Vis spectroscopy and by measuring total organic carbon (TOC). Percentage of removal increased with time, current density, and NaCl concentration. SEC value increased with increased time and current density but it was decreased significantly by NaCl additions (0-29 g L^{-1}). It was found that electro-coagulation treatment of effluents containing phenolic compounds involves complex formation between ferrous/ferric and phenolic compounds present in treated effluent, which has significant impact on the efficiency of the process.

Keywords

Complexation; NaCl; low carbon steel, UV/Vis spectroscopy, total organic carbon (TOC).

Introduction

High amounts of waste water are generated in the Mediterranean area each year during the short periods of the olive oil production process, from October to December. Its direct disposal in nature is not acceptable since it contains dark colour, organic materials, and emulsion oils. Additionally, such waters contain phenolic compounds that have a negative impact on vegetation and microorganisms, and therefore must be treated in order to remove organic and toxic pollutants. The amount of these compounds is usually in the range of 5-25 g L^{-1} depending on climate, variety of olive fruit, cultivation, ripeness at harvest, as well as on extraction process [1,2].

Different methods have been developed for olive mill wastewater (OMW) treatment such as biological treatment [3-5], physico-chemical treatments [6], photocatalytic oxidation [7,8], electrooxidation [9], electrocoagulation [10-13], and a number of combined treatments [14,15]. Phenolic compounds extracted from OMWs could possibly be utilized in the cosmetic, pharmaceutical, or food industries [16], or even as potential source of natural dyes [17]. However, none of treatments used were completely satisfying.

Electrocoagulation is a well-known remediation technique that can be used alone [10-18] or in combination with other techniques [19]. It is a simple, effective, and low cost process, easily adaptable to other systems. Furthermore, in remote areas without electricity, it could be directly powered by a photovoltaic system in order to achieve a self-sustainable unit [20-22].

The goal of this work was to study the influence of NaCl concentration, current density and time on the removal efficiency of phenolic compounds by the electrocoagulation process, as well as to compare specific energy consumption (SEC) of the processes under different experimental conditions. Electrocoagulation was carried out for two different samples of water: model water of mimosa tannin and olive mill wastewater (OMW). Mimosa tannin is a phenolic compound similar to those phenolic compounds present in OMW but it is somewhat more complex. In this work it was used in order to examine the influence of pure phenolic compounds on the removal efficiency of the electrocoagulation process. After detailed analysis of the results obtained in model water, experiments were also carried out in OMW. Electrocoagulation processes are usually carried out without or with only small amounts of NaCl to avoid its presence in discharge water. However, since most olive oil production plants are situated on the Mediterranean coastline, additions of NaCl and the disposal of the treated effluent containing NaCl into the sea is acceptable. Additionally, it is well known that the amount of salt is the crucial parameter for energy consumption in the electrolysis process, and for the feasibility of this technique.

Experimental

All the chemicals used in this research were of analytical grade. Mimosa tannin (Mimosa Central Co-operative Ltd., South Africa) solutions were prepared using bi-distilled water with the addition of an appropriate amount of NaCl.

The olive mill wastewater (OMW) used in this study was obtained from a local olive oil manufacturer (Croatia). It was stored in an open puddle for two months before the sampling. Prior to the experiments the OMW was filtrated to remove suspended solids.

Conductivities and pH values were measured using a conductivity meter (Oakton PCD650) and a pH meter (Radiometer, PHM 220).

At the end of the electrocoagulation process of the mimosa tannin model water, the formed colloids were left during the night to settle and afterwards the sludge was separated from the effluent on a Buchner funnel using a water aspirator. The sludge was dried in open air for three days and then the mass of the sludge was determined. Treated effluents were analysed by different techniques (pH meter, conductivity meter, UV/Vis spectroscopy, and TOC). Before the UV/Vis spectroscopy and TOC analysis were carried out, the solutions was centrifuged using 9000 rpm (Hettich Universal, Mikro 12-24 centrifuge).

At the end of electrocoagulation process of the OMW, the water was immediately filtrated in a Buchner funnel and analysed using the same techniques as in the case of mimosa tannin solution.

The phenolic compounds concentration was measured at the wavelength corresponding to the maximum absorbance using a spectrophotometer (Ocean Optics 200, UV light source Analytical Instrument Systems Inc., Model D 1000 CE) connected to a computer, in 1 cm path-length cells. The equation used to calculate the phenol removal efficiency in the experiments was:

$$R / \% = \frac{\gamma_0 - \gamma}{\gamma_0} 100 \qquad\qquad\qquad (1)$$

where γ_0 and γ are defined as the concentration before and after electrocoagulation process. A similar equation was used to calculate total organic carbon (TOC) removal efficiency ($R_{TOC} / \%$).

The TOC measurements were done on a Shimadzu Analyser TOC V-CSN using the NPOC method.

The electrocoagulation experiments were carried out in a glassy electrolytic cell with dimensions of 7 x 7 x 6 cm. Parallel plate electrodes were immersed in the cell with a working electrode situated between two counter electrodes in order to achieve a good current and potential distribution, and a uniform electrode dissolution. Low carbon steel (composition: 0.06 % C, 0.015 % P, 0.008 % S, 0.007 % Si and 0.35 % Mn) was used for all electrodes. Before the experiment the electrodes were polished using 600 grit emery paper, they were washed with bi-distilled water, and finally, with ethanol. The working electrode had a total immersed area of 10 cm^2, and the counter electrodes had a total immersed area of 20 cm^2. The distance between the electrodes was 1 cm. Before the experiment, the appropriate amount of NaCl was dissolved in the treated solutions. During the experiment constant stirring speeds of 600 rpm and DC Power Supply (Iskra, MA 4165; 1.5 A; 25 V) were used.

All the experiments were carried out at room temperature (23±1 $^\circ$C).

RESULTS AND DISCUSSIONS

Treatment of the model water containing mimosa tannin

The mimosa tannin ($\gamma = 1$ g L^{-1}) solution prepared in 0.5 mol dm^{-3} NaCl has pH 5.1 and $\kappa = 38.3$ mS cm^{-1}. In order to find out the removal efficiency of mimosa tannin over different durations, the process was carried out for 5, 15, and 35 minutes using current densities of 10 mA cm^{-2}. The characteristic voltage of this process was 1.25 V. The influence of the duration of the experiment on the electrocoagulation efficiency is presented in Table 1.

Within the 5 min removal period removal efficiency reached 92.2 %, while further treatment resulted only in its slight increase (35 min, 96.7 %). Energy consumption during the experiment increased proportionally with time since the current density and voltage were constant throughout the experiments.

To explain the mechanism of the electrocoagulation process one must consider the reactions occurring on both electrodes. During the anodic process iron is oxidized and dissolved as Fe^{2+}. Under the experimental conditions used in this work (pH = 5.1) it does not undergo hydrolysis. However, in aerated conditions it is further oxidized by dissolved oxygen to Fe^{3+}, which is susceptible to hydrolysis, resulting in different aqua and hydroxyl complexes, such as Fe(OH)$^{2+}$, Fe(OH)$^+$, Fe(OH)$_3$ under acid conditions, or Fe(OH)$_6^{3-}$ and Fe(OH)$_4^-$ under alkaline conditions [23]. Since the pH values registered before and after electrocoagulation process varied from 5-7, positively charged or neutral particles were expected under the given experimental conditions. Furthermore, negative ions present in water (in the case of NaCl addition it is Cl⁻) will surround a positive charge, forming a diffuse layer which makes the particle neutrally charged. These surface properties make the particles unstable and agglomeration takes place. As a result of agglomeration, particles form flocks precipitating or floating by the bubbles of hydrogen (Figure 1.). The stability of the particles and their agglomeration depends on the type and concentration of ions in the solution. The formed flocks can effectively remove pollutants by adsorption or enmeshment in a precipitate.

The aim of this work was to remove mimosa tannin from model water. It is well known from the literature that dissolved iron in the presence of tannin forms ferrous/ferric tannates. These reactions are pretty complex since both ions, ferrous and ferric, can participate in the reaction, and in addition there is also a possibility of Fe^{3+} reduction by mimosa tannins. According to the author's knowledge, the formation of the complex during phenolic compounds removal by electrocoagulation technique was not considered before, although it might be of a great importance for its progress. It can influence the amount of ferrous/ferric ions required for the coagulation, since Fe^{2+}/Fe^{3+} are consumed by the complex formation. Additionally, it can also impact the mechanism of the reaction because generated complex will be involved in the adsorption or enmeshment by formed flocks instead of mimosa tannin. It could also be important for the electrode reaction kinetics since tannin inhibits dissolution of iron [24,25]. Therefore it can be concluded that complex formation should not be ignored when considering the electrocoagulation process.

Figure 1. *Illustration of flocks precipitating due to gravity, and floating due to the bubbles of gas*

An important parameter of the electrocoagulation process is the effluent pH value at the end of the process. In these experiments pH values decreased for processes conducted over 5 min, while they increased for prolonged treatments (15 and 35 min) (Table 1). The increase of pH values were caused due to the intensive hydrogen evolution at the cathode and the generation of OH⁻ ions. The generated OH⁻ ions were consumed during the hydrolysis of Fe^{3+}, but the overall reaction obviously resulted in the excess of OH⁻ ions. The final pH value depended on the equilibrium of each reaction in the process. Additionally, the increase of pH value can be explained by mimosa tannin removal due to its acidic behaviour.

Table 1. *Results of treatments at different times (j = 10 mA cm^{-2}, U = 1.25 V, γ (NaCl) = 29.22 mol g L^{-1}, κ = 38.3 mS cm^{-1}, pH = 5.1, γ_0(tannin)=1 g L^{-1}, V = 0.1 L).*

T / min	R / %	SEC/ kW h kg^{-1}	m(precipitate) / g	pH(after EC)
5	92.21	0.113	0.143	4.81
15	95.54	0.327	0.197	5.74
35	96.68	0.754	0.263	6.31

Since the concentration of ferrous/ferric ions is dependent on current density, the efficiency of the process is dependent on current density as well. In this work the electrocoagulation process was conducted for 15 min using different current values (Figures 2 and 3). When the current increased from 1 to 10 mA cm^{-2}, removal efficiency increased from 10 to 20 % in 0.58 g L^{-1} NaCl, from 47 to 93 % in 5.84 g L^{-1} NaCl, and from 58 to 96 % in 29.22 mol dm^{-3} NaCl (Figure 2). However, specific energy consumption per mass of mimosa tannin increased more significantly from 0.302 to 7.311 kW h kg^{-1} in 0.58 g L^{-1} NaCl, from 0.034 to 0.469 kW h kg^{-1} in 5.84 g L^{-1} NaCl, and from 0.021 to 0.327 kW h kg^{-1} in 29.22 g L^{-1} NaCl (Table 2). As evident, the highest removal efficiency was obtained at high current density. The obtained results also show that the highest removal efficiency was registered in the presence of high NaCl concentrations, where the lowest specific energy consumption is required. It is supported by Figure 4, where the colour of the solution suggests more difficult precipitation of flocks for small additions of NaCl causing reduced removal efficiency. This is in accordance with theory that coagulation process depends on type and concentration of ions present in solution. It is obvious that optimal process conditions were obtained in the presence of high amounts of NaCl. These results pointed out the importance of the NaCl concentration as a key parameter for an efficient and low cost process.

NaCl's presence is important because of the two effects: (a) it decreased the applied voltage and energy power demand [26] and (b) it changed the ionic strength that affected the coagulation process as evident from Figures 2 and 4. The influence of ionic concentration and zeta potential on the electrocoagulation process was reported previously [27, 28].

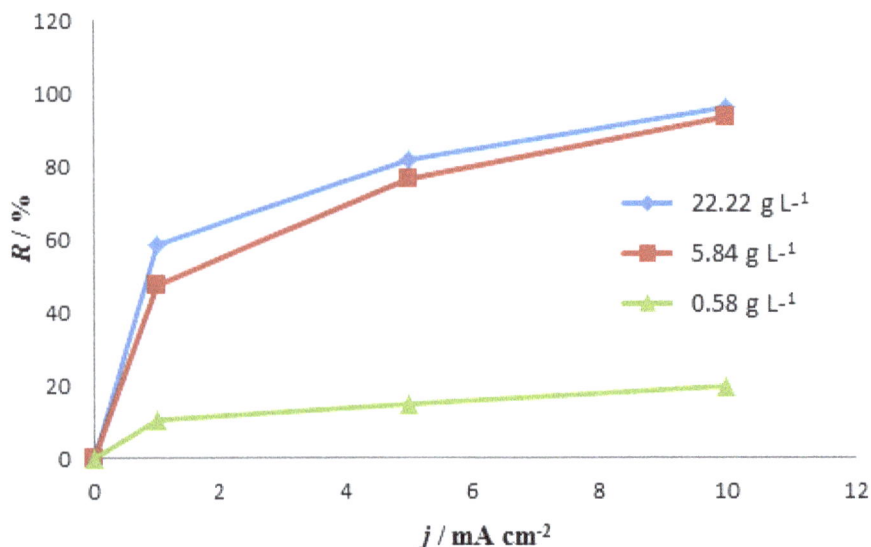

Figure 2. *Influence of NaCl concentration and current density on removal efficiency.*

The main factor influencing energy consumption is applied voltage. The overall voltage is dependent on equilibrium potential difference ($E_{r,k}$-$E_{r,a}$), anode and cathode over-potentials (η_a, η_k), and ohmic potential drop in the solution (η_{IR}) according to the equation (2):

$$|U_{er}| = |(E_{r,k} - E_{r,a})| + \eta_a + |\eta_k| + \eta_{IR} \qquad (2)$$

Ohmic potential drop in the solution is dependent on cell configuration, electrode area (A / m²), and the distance between electrodes (d / m), as well on the conductivity of solution (κ / S cm⁻¹):

$$\eta_{IR} = I\left(\frac{d}{A\kappa}\right) \qquad (3)$$

Table 2. Results of treatments at different NaCl concentration and current density
(t = 15 min, pH = 5.1, γ(tannin) = 1 g L⁻¹, V = 0.1 L).

γ (NaCl) / g L⁻¹	J / mA cm⁻²	U / V	SEC / kW h kg⁻¹	m(precipitate) / g
29.22 (κ = 38.3 mS cm⁻¹)	10	1.25	0.327	0.197
	5	0.95	0.146	0.134
	1	0.50	0.021	0.103
5.84 (κ = 10.11 mS cm⁻¹)	10	1.75	0.469	0.140
	5	1.15	0.188	0.128
	1	0.65	0.034	0.094
0.58 (κ = 0.98 mS cm⁻¹)	10	5.75	7.311	0.080
	5	3.65	3.061	0.044
	1	1.25	0.302	0.038

Figure 3. Results of treatments at different current densities in the case of mimosa tannin:
(a) 10 mA cm⁻²; **(b)** 5 mA cm⁻²; **(c)** 1 mA cm⁻² (t = 15 min; 29.22 g L⁻¹NaCl).

Figure 4. Results of treatments at different NaCl concentrations in the case of mimosa tannin:
(a) 29.22 g L⁻¹ NaCl; **(b)** 5.84 g L⁻¹NaCl; **(c)** 0.58 g L⁻¹NaCl (j = 10 mA cm⁻²; t = 15).

SEC is dependent on current value (I), applied voltage (U) and time (t) and it is expressed as energy consumption per mass of removed tannin:

$$SEC = \frac{IUt}{\left(\gamma\left(tannin\right)_0 - \gamma\left(tannin\right)\right)V} \tag{4}$$

where γ is the mass concentration of tannin or phenolic compound, V is volume of treated solution.

Another important parameter for SEC is the distance between the electrodes, which in most of the previous reports ranged from 0.3-3.0 cm, while the distance in this work was 1 cm. A small distance is preferable to decrease potential drop, but the electrodes should be adequately separated in order to enable unhindered movement of flocks between them.

Conductivity of the solution i.e. salt concentration also plays important role for SEC value and according to our knowledge its value was quite different in different reports.

Kobaya et al. [29] treated textile wastewater by electrocoagulation process and it was shown that an addition of NaCl (κ = 1000-4000 μS cm^{-1}) did not influence process efficiency but energy consumption decreased with increased wastewater conductivity (2.2-0.75 kW h kg^{-1} (COD)). Sengil et al. [30] have used electrocoagulation for decolourization of Reactive red and it was found that small additions (0.5-2.0 g L^{-1}) of NaCl increase efficiency while further addition did not have any impact. The optimal conditions were found to be 2.3 g L^{-1} NaCl and 4.54 kW h kg^{-1} (dye). B. K. Nandi et al. [31] varied NaCl concentration from 0.1-1.0 g L^{-1} and it was found that efficiency had increased from 97-100 % and energy consumption had decreased from 17-3 kW h kg^{-1} (Fe). X. Chen et al. [32] separated pollutants from restaurant wastewater by electrocoagulation process without the addition of NaCl when energy consumption was in the range of 0.2-1.4 kW h m^{-3} depending on the solution conductivity that varied from 770-227 μS cm^{-1}. The additions of NaCl changed conductivity from 443-2850 μS cm^{-1} and energy consumption as well, from 0.32-0.29 kW h m^{-3}; however, it did not change the efficiency of the process.

From the previous results it follows that NaCl concentration can influence specific energy consumption drastically, which will have an impact on operating cost. However, the dependence of removal efficiency on NaCl concentration is not completely clear and it depends on the type of pollutant and its concentration.

The results of this paper confirm that NaCl addition decreases specific energy consumption in accordance with the previous results. Furthermore, it was shown that efficiency of the phenolic compound removal can also be improved considerably by the addition of NaCl in the range from 0.58 g L^{-1} to 29.22 g L^{-1}.

Treatment of OMW

The starting OMW solution had the following characteristics: pH 5.37, concentration of phenolic compounds, γ_0 = 0.613 g L^{-1} (mimosa tannin equivalent), and the TOC value was 1376 mg L^{-1}. These values were somewhat lower in comparison to the values frequently found in the literature [10-12]. This can be explained by the fact that the OMW was kept in an open puddle for 2 months.

During OMW treatment by electrocoagulation process in previous investigations the solution as received was used or it was diluted with water. The conductivity of the pure OMW sample was 11 mS cm^{-1} [10-12] and for diluted OMW (1:5) conductivity was 3.6 S cm^{-1} [13]. The SEC value obtained during the OMW treatment was found to be 4 kW h kg^{-1} (COD) [13] or 20-300 kW h m^{-3} (volume of treated solution) [12], which were quite similar considering the characteristic COD value for OMW. Also, it was shown that small additions of NaCl improve removal efficiency while additions higher than 2 g L^{-1} decrease removal efficiency. Energy consumption has decreased upon NaCl addition.

From the results of the treatment of mimosa tannin, the current density of 10 mA cm^{-2} was chosen for the OMW treatment. The process was carried out with different additions of NaCl from 0-20 g L^{-1} during 35 or 60 min. Depending on NaCl addition conductivity of the solutions was changed from the value similar to the previously reported values (2.3 S cm^{-1}) to the value higher than previously reported (23.7 S cm^{-1}). The SEC value was changed from 8.5-1.6 kW h kg^{-1} (mass of phenolic compounds) during 35 min or it was changed from 8.2-2.6 kW h kg^{-1} (mass of phenolic compounds) during 60 min.

Similarly as in the case of model water, addition of NaCl had positive impact on removal efficiency (Figures 5 and 6) and energy consumption (Table 3). Furthermore, better efficiency was obtained by prolonged process time while SEC was not increased significantly. At the end of the electrocoagulation process pH value was close to 7, which was acceptable for discharge, while conductivity increased only slightly.

Table 3. Results of treatments at different NaCl concentration and process times
(j= 10 mA cm^{-2}. pH = 5.37. V= 0.1 L, γ_0 = 0.613 g L^{-1}).

t/ min	γ(NaCl) / g L^{-1}	U / V	SEC/ kW h kg^{-1}	$pH_{after\ EC}$	κ(OMW)$_{before\ EC}$/ mS cm^{-1}	κ(OMW)$_{after\ EC}$/ mS cm^{-1}
35	0	3.8	8.492	6.65	2.32	3.43
	5	1.9	2.677	6.85	8.23	9.48
	10	1.5	1.998	6.88	13.77	14.75
	20	1.4	1.604	6.9	23.72	24.79
60	0	3.9	8.226	6.7	2.32	3.26
	5	2.1	4.26	6.81	5.99	6.55
	10	1.6	2.941	6.94	10.12	10.98
	20	1.5	2.562	6.98	20.44	21.75

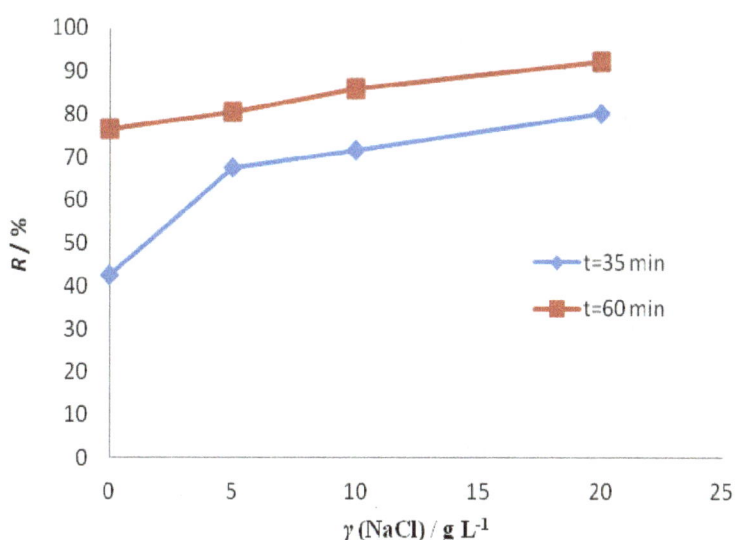

Figure 5. Dependence of removal efficiency of phenolic compound in OMW treated effluent on NaCl concentration and process time.

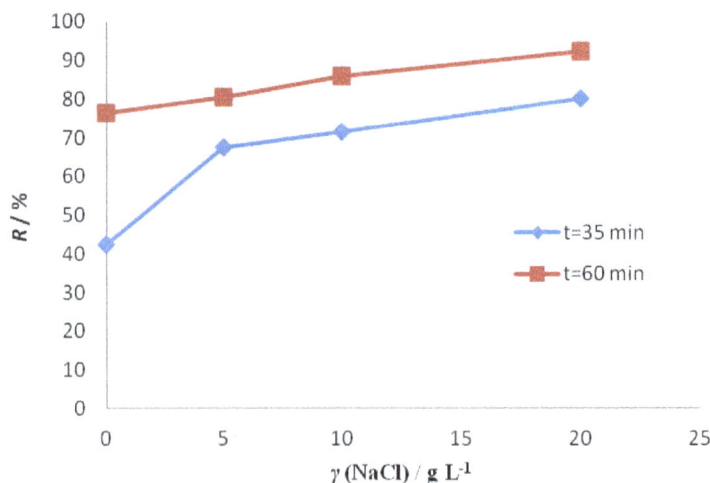

Figure 6. *Dependence of removal efficiency of TOC in OMW treated effluent on NaCl concentration and process time.*

From Figures 5 and 6 it is evident that the removal of overall organic loading (TOC) is lower (30 - 70 %) in comparison to phenolic compounds (40 - 90 %). Better efficiency in the case of phenolic compounds could be the consequence of complex formation between ferrous/ferric and phenolic compounds present in the OMW. It is also evident that removal efficiency of phenolic compounds in OMW was lower compared to the removal efficiency of mimosa tannin, although longer times were used (Figures 2 and 5). It is not surprising since this solution, apart from the phenolic compounds, contains some other constituents such as oil, sugar, and pulp suspension [2]. Therefore, the capacity of produced sludge for phenolic compounds removal is reduced in the case of OMW compared to the model water.

Conclusions

The results obtained in this paper show that it is possible to obtain high removal efficiency of mimosa tannin and phenolic compounds from OMW by electrocoagulation process. In the case of mimosa tannin, electrocoagulation was able to reduce the phenolic content up to 96 %, while in the case of OMW electrocoagulation wasa able to reduce the phenolic content up to 92 %. The percentage of removal was increased with increased time, current density, and NaCl concentration. Apart from the increasing removal efficiency of the process, an improvement in energy demand was also obtained with the addition of NaCl. Therefore, it can be concluded that an addition of NaCl can significantly improve the electrocoagulation process. Furthermore, additions of NaCl and the disposal of treated effluent containing NaCl are acceptable for the production plants located close to the coast. At the end of process pH value was close to 7, which is acceptable for discharging.

It was shown that complex formation between ferrous/ferric and phenolic compounds present in treated effluent could change the efficiency of the process. Thus, due to the complexation, removal of phenolic compounds was higher in comparison to removal of overall organic loading (TOC). It was also observed that the removal efficiency of mimosa tannin is higher compared to the removal efficiency of phenolic compounds from OMW although longer times were used. It is explained by the decreased capacity of produced flocks for phenolic compounds removal, in the case of OMW, due to the presence of other organic constituents.

Acknowledgements: *Financial support by Ministry of Science, Education and Sports of Republic of Croatia (project 125-1252973-2576) is gratefully acknowledged. The authors express their - gratitude to Višnja Pavić for providing mimosa tannin.*

References

[1] http://www.agbiolab.com/files/agbiolab_Polyphenols.pdf

[2] F. Federici, *Pomologia Croatica* **12** (2006) 15-27.

[3] D. Quaratino, A. D'Annibale, F. Federici, C.Cereti, F. Rossini, M. Fenice, *Chemosphere* **66** (2007) 1627–1633.

[4] T. Landeka Dragičević, M. Zanoški Hren, M. Gmajnić, S. Pelko, D. Kungulovski, I. Kungulovski, D. Čvek, J. Frece, K. Markov, F. Delaš, *Archives of Industrial Hygiene and Toxicology* **61** (2010) 399-405.

[5] P. Paraskeva, E. Diamadopoulos, *Journal of* Chemical *Technology and Biotechnology* **81** (2006) 1475–1485.

[6] A.C. Barbera , C. Maucieri , A. Ioppolo, M. Milani, V. Cavallaro, *Water Research* **52** (2014) 275-281.

[7] I. Michael, A. Panagi, L.A. Ioannou, Z. Frontistis, D. Fatta-Kassinos, *Water Research* **60** (2014) 28-40.

[8] E. Chatzisymeon, E. Stypas, S. Bousios, N.P. Xekoukoulotakis, D. Mantzavinos, *Journal of Hazardous Materials* **154** (2008) 1090-1097.

[9] U.T. Un, U. Altay, A. S. Koparal, U.B. Ogutveren, *Chemical Engineering Journal* **139** (2008) 445–452.

[10] U . T. Ün, S. Ugur , A.S. Koparal , U. B. Öğütveren, *Separation and purification technology* **52** (2006) 136-141.

[11] N. Adhoum, L. Monser, *Chemical Engineering and Processing* **43** (2004) 1281-1287.

[12] H. Inan, A. Dimoglo, H. Dimsek, M. Karpuzcu, *Separation and purification technology* **36** (2004) 23-31.

[13] F. Hanafi, O. Assobhei, M. Mountadar, Journal *of Hazardous Materuials* **174** (2010) 807-812.

[14] S. Khoufi, F. Aloui, S. Sayadi, *Water Research* **40** (2006) 2007 – 2016.

[15] J.M. Ochando-Pulidoa, G. Hodaifab, M.D. Victor-Ortegaa,S. Rodriguez-Vivesa, A. Martinez-Ferez, *Journal of Hazardous Materials* **263P** (2013) 158– 167.

[16] J H. Zbakh, A. El Abbassi, *Journal of* Functional *Foods* **4** (2012) 53-65

[17] N. Meksi, W. Haddar, S. Hammami, M.F. Mhenni, *Industrial Crops and Products* **40** (2012) 103– 109.

[18] S. Zodi, O. Potier, C. Michon, H. Poirot, G. Valentin, J. P. Leclerc, F. Lapicque, *Journal of Electrochemical Science and Engineering* **1** (2011) 55-65.

[19] A. Aouni, C. Fersi, M. B. Sik Ali, M. Dhahbi, Journal *of Hazardous Materials* **168** (2009) 868–874.

[20] J. M. Ortiz, E. Expósito, F. Gallud, V. García-García, V. Montiel, A. Aldaz, *Desalination* **208** (2007) 89–100.

[21] D. Valero, J. M. Ortiz, E. Expósito, V. Montiel, A. Aldaz, *Solar Energy Materials & Solar Cell* **92** (2008) 291-297.

[22] R. García-Valverde, N. Espinosa, A. Urbina, International *Journal of HydrogenEenergy* **36** (2011) 10574-10586.

[23] M. Yousuf, A. Mollah, R. Schennach, J. R. Parga, D. L. Cocke, *Journal of Hazardous Materials* **B84** (2001) 29-41.

[24] S. Yahya, A. M. Shah, A. A. Rahim, N. H. A. Aziz, R. Roslan, *Journal of Physical Science* **19** (2008) 31-41.

[25] S. Martinez, *Materials Chemistry and Physics* **77** (2002) 97–102.

[26] A. N. Subba Rao and V. T. Venkatarangaiah, *Journal of Electrochemical Science and Engineering* **3** (2013) 167-184.

[27] C. Barrbera-Díaz, B. Frontana-Uribe, B. Bilyeu, *Chemosphere* **105** (2014) 160-164.

[28] M. Vepsäläinen, M. Pulliainen, M. Sillanpää, *Separation and Purification Technology* **99** (2012) 20-27.

[29] M. Kobya, O. T. Can, M. Bayramoglu, *Journal of Hazardous Materials* **B100** (2003) 163-178.

[30] I. A. Şengil, M. Özacar, B. Ömürlü, *Chemical* and *Biochemical Engineering Quarterly* **18** (2004) 391-401.

[31] B. K. Nandi, S. Pateol, *Arabian Journal of Chemistry,* In Press, Corrected Proof, DOI: 10.1016/j.arabjc.2013.11.032

[32] X. Chen, G. Chen, P. L. Yue, *Separation and Purification Technology* **19** (2000) 65-76.

11

Effect of corrosion on flexural bond strength

Akshatha Shetty[✉], Katta Venkataramana and K. S. Babu Narayan

Department of Civil Engineering, NITK Surathkal-575025, India

[✉]*Corresponding author*

Abstract

Corrosion is one of the main causes affecting durability of structures. Corrosion effects on structures cannot be ignored and replaced. To understand the performance of structures there is a need to study the rate at which different corrosion levels occur. Hence the present investigation has been taken up to study the behaviour of NBS (National Bureau of Standard) beam specimens made up of Ordinary Portland Cement (OPC) and Portland Pozzolona Cement (PPC) concrete matrix were subjected to accelerated corrosion for different corrosion levels of 2.5 % to 10 % at 2.5 % interval. Results are compared with those for control beam specimen. It is observed that bond stress value decreases with the increase in corrosion levels. Also corrosion leads to the decline of load carrying capacity.

Keywords

OPC concrete; PPC concrete; Corrosion level; NBS beam; Strain; Corrosion current density

Introduction

Reinforcement corrosion has been identified as being the predominant deterioration mechanism for reinforced concrete structures, which seriously affects the serviceability and safety of structures.

Chloride ions from the external environment diffuse through concrete to the steel surface, leading to the depassivation of the protective layer and ultimately to the initiation of reinforcement corrosion. Corrosion consumes iron of the reinforcing bar progressively thereby reducing the cross sectional area. Corrosion increases its volume 2 to 6 times than that of the original steel; it causes volume expansion developing tensile stresses in concrete [1].

The corrosion of rebar in concrete is generally considered as an electrochemical process [2]. With attention of researchers focusing towards the prediction of the residual life of reinforced concrete structures affected by reinforcement corrosion, the use of electrochemical techniques for the determination of relevant parameters in this regard becomes a major area of study.

Therefore the electrochemical techniques are widely used for the study of rebar corrosion in laboratories together with their application to real life structures [3].

The bond between steel and concrete is the mechanism which allows for effective force transfer. The load applied to the concrete member is transferred to the reinforcing steel through bond [4]. Bond of deformed bars is developed mainly by the bearing pressure of the bar ribs against the concrete [5]. One of the most important prerequisites of reinforced concrete construction is adequate bond between the reinforcement and the concrete. Hence present study aims to investigate the effect of corrosion on performance of Reinforced concrete beams, in particular on the bond strength and load carrying characteristics and deflection characteristics and maximum crack width. Following are the main highlights of the study:

1. Accelerated corrosion using impressed current technique was adopted for achieving the desired corrosion levels.
2. NBS RC beams have been cast and tested for bond strength by two-point loading flexural strength.
3. OPC and PPC concrete were used for preparing the beam specimens.

Experimental

Details of experiments, materials used, and method of testing are explained below.

Materials

Materials used for the experimental investigation are tested as per code provisions:

Cement - Ordinary portland cement and Portland pozzolona cement were used. Cements used for the study are tested as per IS: 8112-1989 [6] for OPC and IS 1489 (Part-I):1991 [7] for PPC recommendations. Setting time, specific gravity and compressive strength of results are tabulated in Table 1 and Table 2 respectively.

Table 1. Test results on Characteristics of OPC

Sl No	Test Parameters	Results	As per IS 8112:1989 (Specifications of 43 Grade OPC)
1	Initial setting and final setting time	75 min and 260 min	Not less than 30 min. and not more than 600 min
2	Specific gravity	3.1	
3	Compressive strength:		
	3 Days	24.09 N/mm^2	Not less than 23 N/mm^2
	7 Days	34.48 N/mm^2	Not less than 33 N/mm^2
	28 Days	46.85 N/mm^2	Not less than 43 N/mm^2

Fine Aggregate - Physical tests on fine aggregates were conducted as per IS 383-1970 [8] and specific gravity, water absorption and moisture content and grading results are presented in Table 3.

Coarse Aggregate - Size of aggregate used was 20 mm downsize and 12.5 mm down size angular type coarse aggregate. Physical tests on aggregates were conducted as per IS 383-1970 [8]. Specific gravity, water absorption and moisture content results are tabulated in Table 4.

Table 2. Test results on Characteristics of PPC

Sl No	Test parameters	Results	As per IS 1489 (Part-I): 1991 (Specifications of Portland Pozzolana cement)
1	Initial setting and final setting time	76 min and 270 min	Not less than 30 min. and not more than 600 min
2	Specific gravity	2.91	
3	Compressive strength:		
	3 Days	16.64 N/mm^2	Not less than 16 N/mm^2
	7 Days	26.54 N/mm^2	Not less than 22 N/mm^2
	28 Days	39.84 N/mm^2	Not less than 33 N/mm^2

Table 3. Test result of fine aggregate used for concrete mix

Sl No	Test Parameters	Results
1	Specific gravity	2.6
2	Water absorption	2.0 %
3	Moisture content	5.0 %
4	Grading	Zone I

Table 4. Test result of 20mm down size used for concrete mix

Sl No	Test Parameters	Results
1	Specific gravity	2.8
2	Water absorption	0.5 %
3	Moisture content	Nil
4	Shape	Angular

Reinforcing steel - Tensile strength of reinforcing steel bar was tested using Universal Testing Machine (UTM). Stress-strain curves for 25 mm Thermo Mechanically Treated (TMT) Fe-415 reinforcing steel bar was obtained by plotting tension test data. Typical stress-strain curve for 25 mm bar is shown in Figure 1. The yield strength and ultimate strength are 485 N/mm^2 and 589 N/mm^2 respectively.

Mix design details

According to the codal recommendation of IS 456-2000 [9] minimum concrete grade to be adopted in coastal environment is M30 and maximum water cement ratio of 0.45. Target strength of 30 N/mm^2 was considered for the study. To Control the proportion of water in freshly mixed concrete is to specify an upper limit for "slump". Hence a slump range of 50-60 mm was selected for the present study. Mix design calculations were made as per IS 10262-2009 [10]. Test results of materials were used in the calculation of determination of mix proportions. After several trials the mix proportion of 1:1.77:2.87 was achieved with an addition of 2 ml/kg of commercially available chemical admixture for both OPC and PPC concrete.

Preparation of test specimens

For the present study National Bureau of Standard (NBS) beam specimens of size 2.44 × 0.457 × 0.203 m (Figure 2) was used [11]. Before placing of concrete in beam molds, one blue color multi-strand copper wire of 4 cm^2 cross sectional area was connected at one end of rebar, and soldered to reinforcement bar and also covered with M-seal to prevent corrosion at that area. This wire is protruded to the surface level to induce electric current. Similar steps were followed at the other end and one black color, multi-strand copper wire of 2 cm^2 cross sectional area was connected to reinforcement bar, which helps in the monitoring process of corrosion rate.

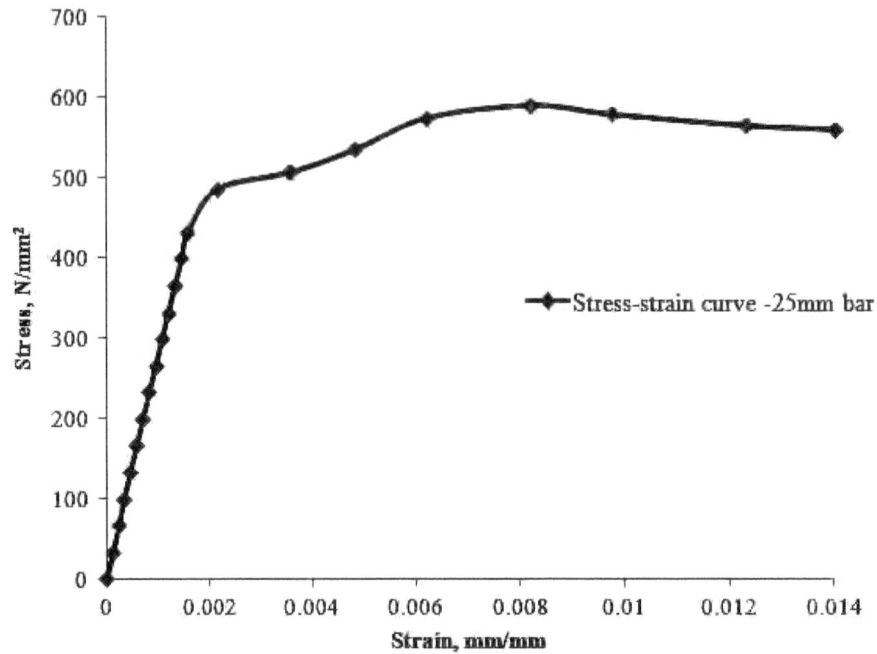

Figure 1. Stress-strain curve for 25 mm diameter TMT Fe-415 reinforcing steel bar

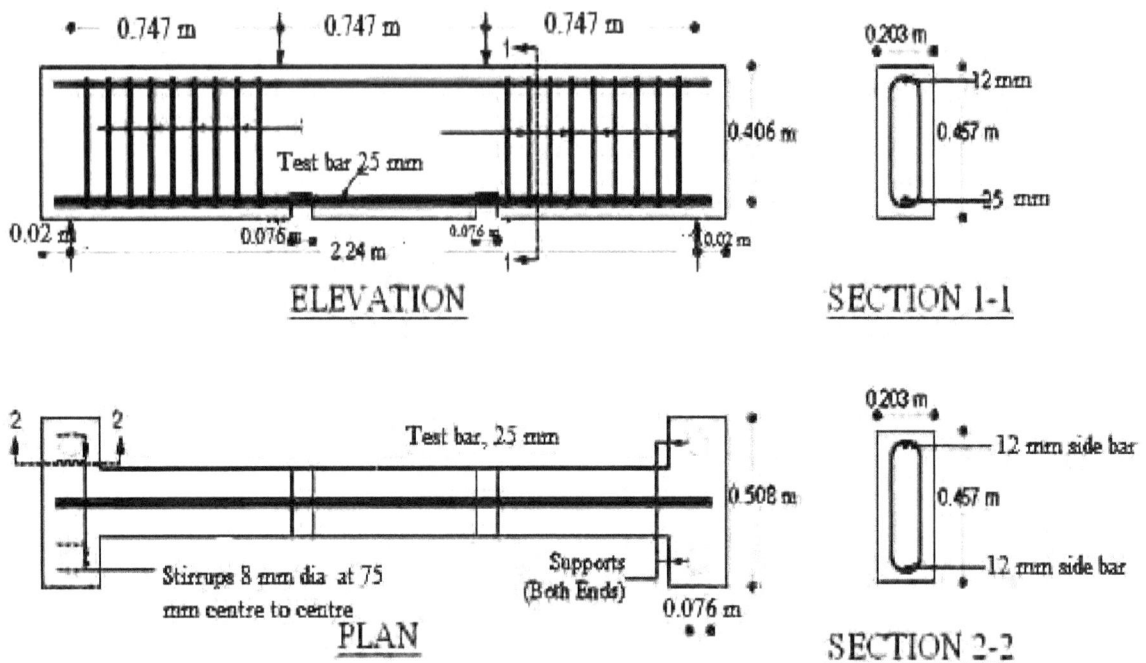

Figure 2. Reinforcement details of NBS beam specimen [11]

Accelerated corrosion technique

Under natural conditions the process of reinforcement corrosion is very slow; so laboratory studies need an acceleration of corrosion process to achieve a short test period. This can be accomplished by applying a constant potential or an electric current of constant magnitude to the embedded steel [12]. Hence accelerated corrosion technique was used in the present study to achieve different degree of corrosion levels.

After curing of beam specimens for 28 days, specimens were lifted and shifted to the corrosion tank to induce desired corrosion levels. Electrochemical corrosion technique was used to accelerate the corrosion of steel bars embedded in beam specimens. Specimens were partially immersed in a 5% NaCl solution for duration of 8 days; direction of current was arranged such that, rebars embedded inside the concrete specimens served as anode. Steel plate which was placed along the length of beam functions as cathode. Current required achieving different corrosion levels can be obtained using Faraday's law (Eq. 3) [13]. Based on the calculation amount of 2.5 to 10 A current at the variation of 2.5 A was applied to obtain the required corrosion level *i.e.* 2.5 to 10 % at the variation of 2.5 % respectively. For each trial, three specimens were considered. A Schematic representation of corrosion test set-up used for accelerated corrosion process is shown in Figure 3.

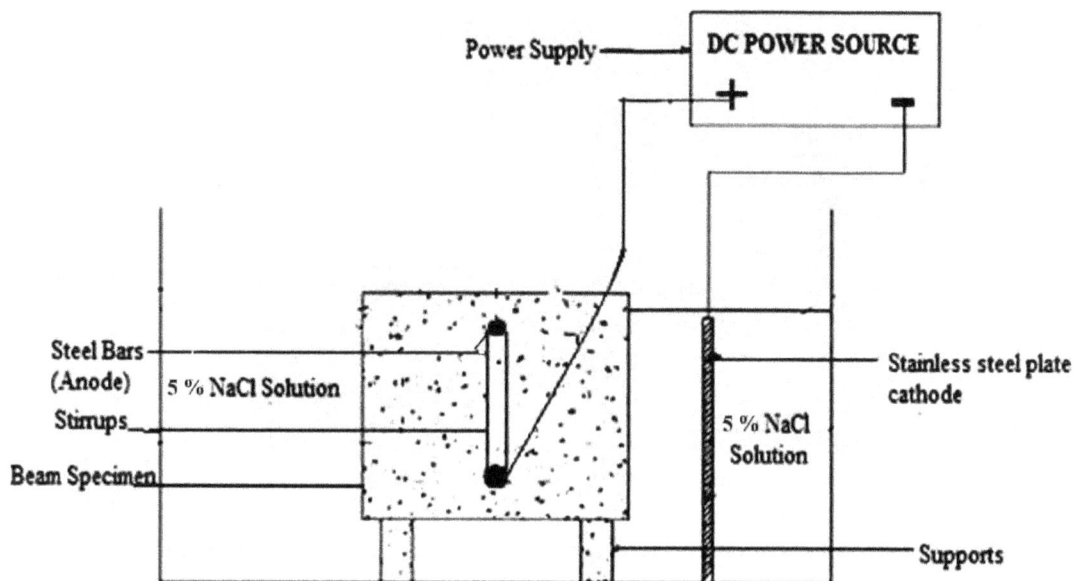

Figure 3. Schematic representation of accelerated corrosion of beam specimen

Calculation of amount of current required to obtain different corrosion levels

From Faraday's law,

$$i_{corr} = \frac{(w_i - w_f)F}{\pi DLWT} \tag{1}$$

$$\rho = \frac{(w_i - w_f)}{w_i} 100 \tag{2}$$

Eq (2) in Eq. (1)

$$i_{app} = i_{corr} = \frac{\rho w_i F}{100 \pi DLWT} \tag{3}$$

where i_{corr} = corrosion current density; i_{app} = applied current; ρ = degree of corrosion W_i = Initial weight of steel (20,000 g); F = 96 487 As; D = bar diameter; L = length of bar; W =equivalent weight of steel (27.925 g); T = time in seconds.

Corrosion rate measurements

After completion of accelerated corrosion, corrosion rate was measured with Applied Corrosion Monitoring (ACM) instrument (Figure 4) based on linear polarization resistance (LPR) method to assure the achieved desired degree of corrosion levels in beam specimens after inducing corrosion for a particular duration.

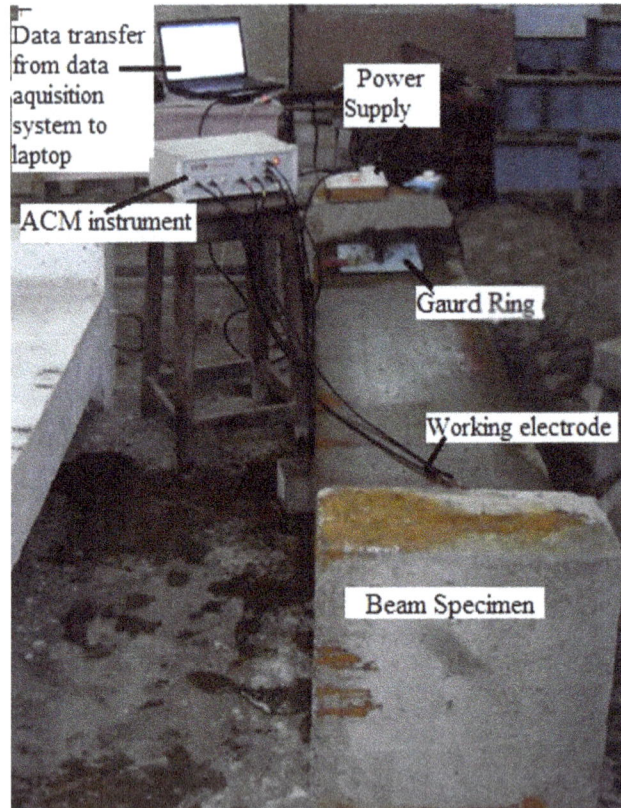

Figure 4. *Corrosion monitoring set up*

Corrosion current density was calculated by using the Stern-Geary formula [14].

$$i_{corr} = \frac{B}{R_p} \tag{4}$$

where, i_{corr} = corrosion current density, $\mu A/cm^2$; R_p= polarization resistance, $k\Omega\ cm^2$; B = 26 mV (for steel in active condition this value is normally used) [15].

Test setup used for flexural bond study

Test set up used for the present study is shown in Figure 5a. Beam specimens were tested under two point loading condition. The load was applied at 15 kN increments. Proving ring of 50 tonne capacity was used to note the applied load.

Strain value recordings have been done by using demec gauges at every load interval. Positions of demec targets have been shown in Figure 5b. Maximum crack width was measured using Crack microscope (Figure 6). Deflection recordings were done by using dial gauges. Dial gauges were indicated in Figure 7.

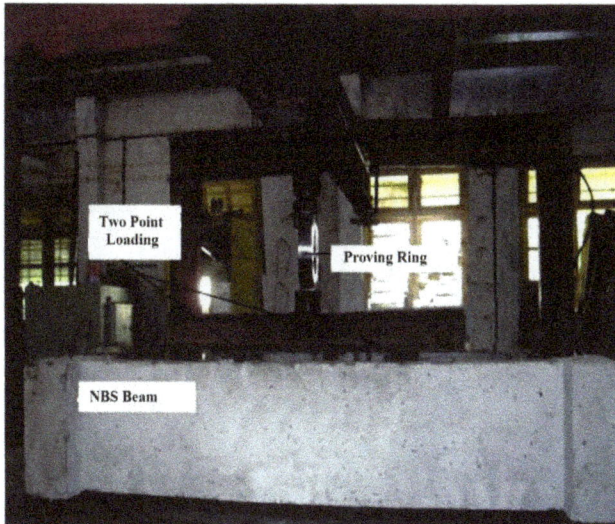

Figure 5(a). *Test set up of NBS beam Specimen*

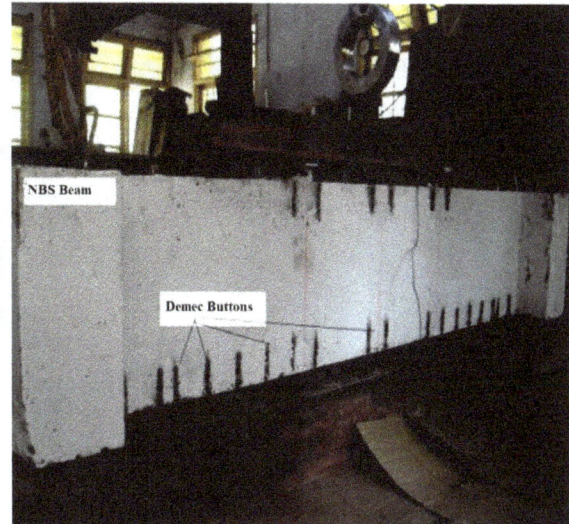

Figure 5(b). *Demec buttons were at 100 mm c/c*

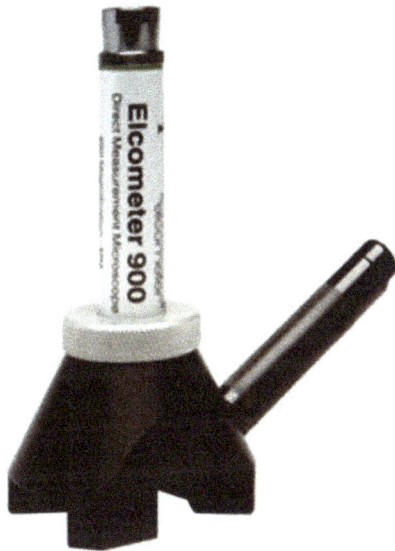

Figure 6. *Concrete crack microscope*

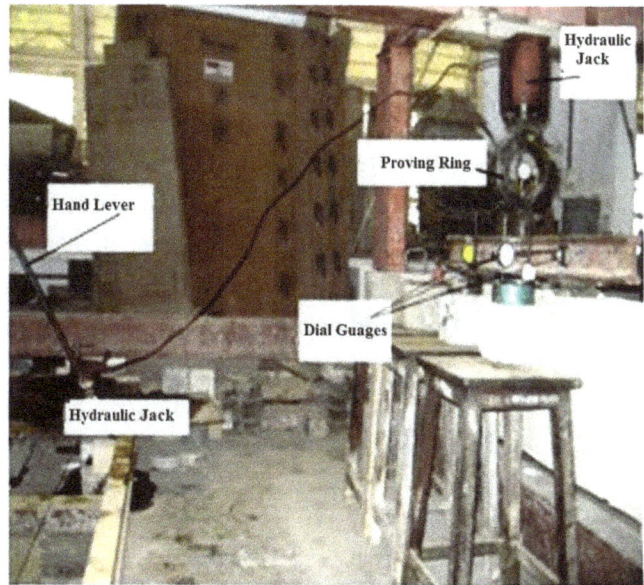

Figure 7. *Positions of dial gauges*

Determination of bond stress

Average bond stress values were obtained from Eq. (6).

$$\tau_{bd} = \frac{\varphi_1 f_s}{4 l_d} \tag{6}$$

where, diameter of bar

$$\varphi_1 = \varphi \sqrt{1 - \frac{p}{100}} \tag{7}$$

τ_{bd} = Average bond stress, N/mm^2

ϕ = initial diameter 25 mm

ϕ_1 = Reduced diameter values are presented in Table 5.

p = Weight loss in percentage

l_d = Embedment length of the bar (747 mm) from the test setup

f_s = Steel stress values have been obtained for initial and final strain values (slip region) of different corrosion levels from stress corresponding to strains at that load level.

Table 5. *Reduced bar diameter for different levels of corrosion*

Corrosion level, %	Bar diameter, mm
0	25.00
2.5	24.69
5	24.37
7.5	24.04
10	23.72

Results and Discussion

Compressive strength of control cube after 28 days of curing was 35 N/mm^2 and 33 N/mm^2 for OPC and PPC concrete respectively.

Measurement of corrosion current density

Corrosion rate is measured in terms of corrosion current density, i_{corr}, and is a quantitative index, which represents an overall estimate of the corrosion attack on reinforcement.

Corrosion current density values were calculated for each specimen from ACM instrument for different grids and average value was considered for different corrosion levels to calculate the weight loss, %. Obtained corrosion levels or weight loss for applied current are shown in Figure 8. It is seen that corrosion levels increases linearly with the increase in the applied current. From Figure 8 corrosion levels for the applied current, A, can be calculated as:

$$y = 0.96x - 0.22 \quad R^2 = 0.995 \quad \text{(PPC Concrete)} \tag{8a}$$

$$y = 0.98x - 0.54 \quad R^2 = 0.998 \quad \text{(OPC Concrete)} \tag{8b}$$

where; x = applied current, A and y = obtained corrosion level, %.

Experimental Investigation on ultimate load carrying capacity and bond stress of RC member

In the present study, 3 control specimens and 12 corroded specimens (3 sets each for 2.5 %, to 10 % at 2.5 % variation of corrosion) were considered for OPC and PPC concrete beam specimens.

Ultimate load carrying capacity, deflection and crack propagation of NBS beams

Effect of corrosion on ultimate load carrying capacity is shown in Figure 9. As the degree of corrosion level increases load carrying capacity decreases (Figure 9). It is also observed that for every percentage increase in corrosion level there is about 1.6 % decrease in load carrying capacity.

From Figure 10 it is noticed that as the corrosion level increases deflection value increases. This is due to as the corrosion level increases cross sectional area reduces and leads to reduction in stiffness of reinforcement.

From Figure 11 it is observed that at low corrosion levels (0 to 5%) crack width is less but as the corrosion level increases (7.5 % to 10 %) crack width increases rapidly. From Figure 12 it is seen that as the load level increases strain value increases linearly in the initial stage.

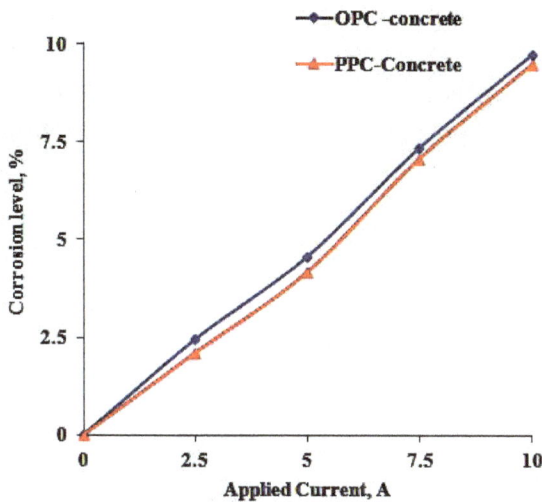

Figure 8. *Variation of corrosion levels with Applied current*

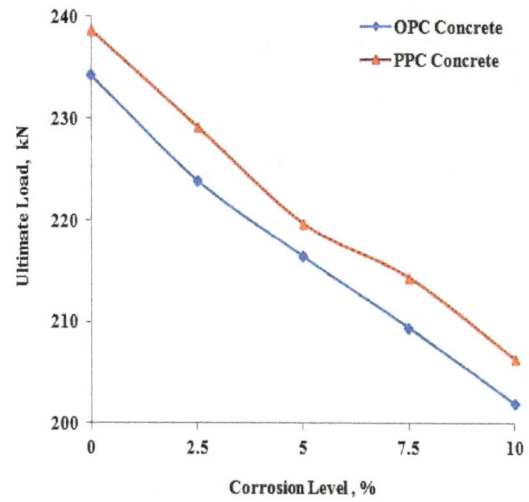

Figure 9. *Effect of corrosion on ultimate load carrying capacity*

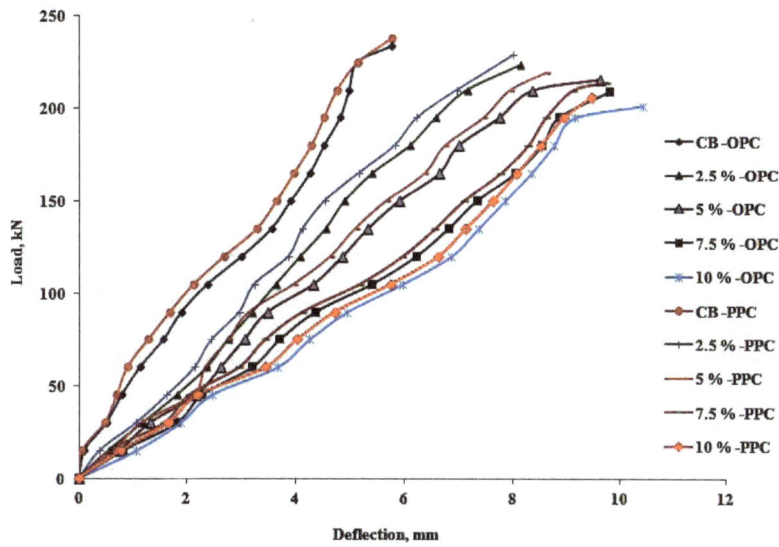

Figure 10. *Effect of corrosion levels on central deflection of NBS beam*

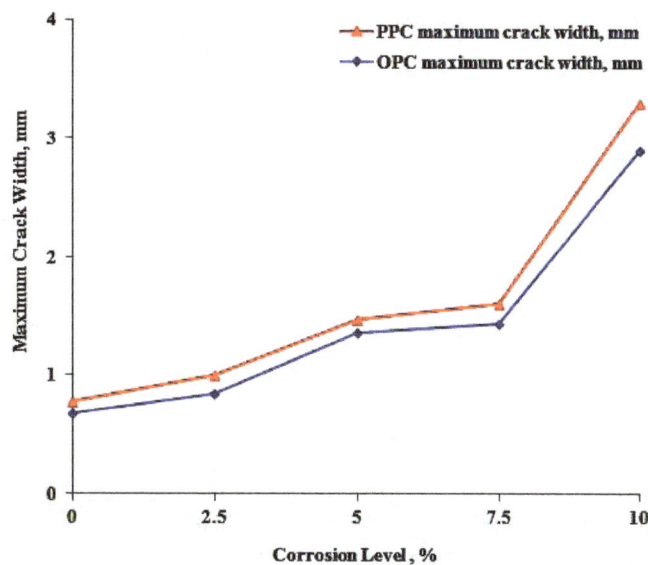

Figure 11. *Effect of corrosion levels on maximum crack width of NBS beam*

Then at higher corrosion levels, rate of increase of strain is higher for the same increment of load level compared to lower levels of corrosion. Control beam specimen performs better at increased corrosion levels. It is also observed that there is a sudden increase in strain values observed for the applied load interval, in all beam specimens (Figures 12 to 15). Bond characteristic is dependent on characteristics of steel at interface and not in the body of reinforcement. From the stress strain curves it is clear that the sudden increase in strain levels without increase in stress is indicative of slip (sudden increases in strain value is much lesser than the yield strength of 25 mm diameter) characteristics owing to drop in bond. Higher the accelerated corrosion at surface lower is the stress level at which such slip occurs.

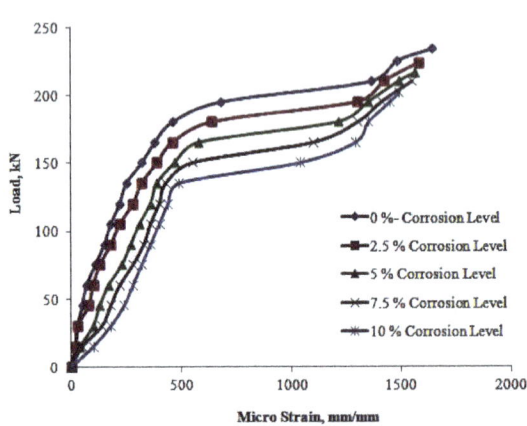

Figure 12. Effect of different levels of corrosion on stress strain behavior of OPC Concrete Beam specimens at left side loading Point (L); where, TRL-Tension side Reinforcement level

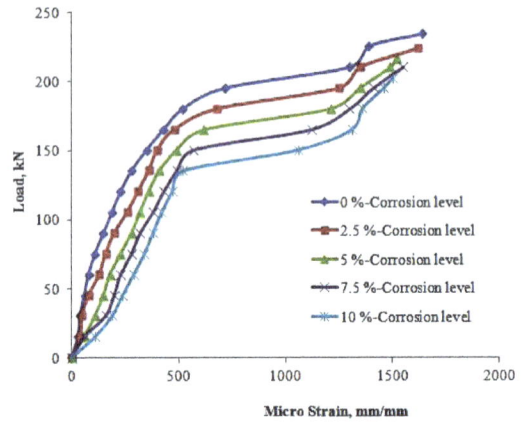

Figure 13. Effect of different levels of corrosion on stress strain behavior of OPC Concrete Beam specimens at right side loading Point (R); where, TRL-Tension side Reinforcement level

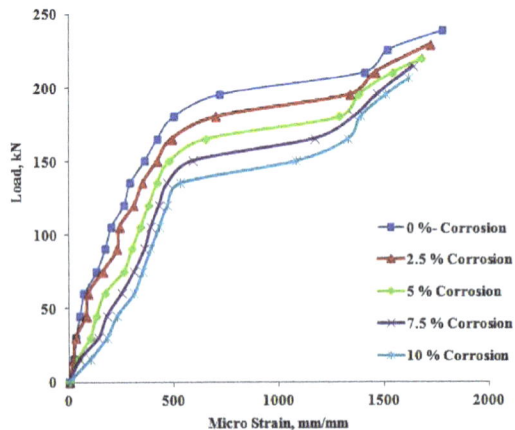

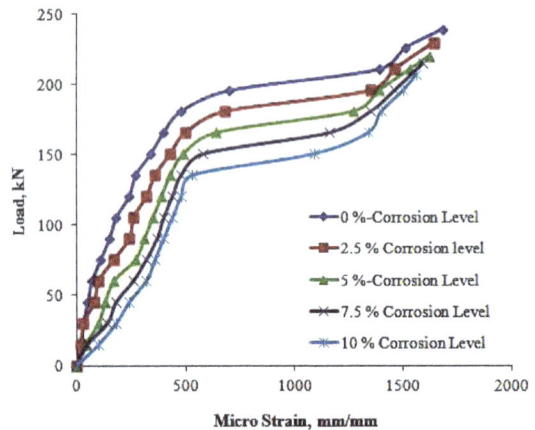

Figure 14. Effect of different levels of corrosion on stress strain behavior of PPC Concrete Beam specimens at right side loading Point (R); where, TRL-Tension side Reinforcement level

Figure 15. Effect of different levels of corrosion on stress strain behavior of PPC Concrete Beam specimens at left side loading Point (L); where, TRL-Tension side Reinforcement level

Bar stress and bond stress performance of NBS beam

From Table 6 and Table 7 it is observed that as the corrosion level increases bond stress value decreases. Percentage reduction in bond stress for different levels of corrosion *i.e.* 2.5 %, 5 %, 7.5 % and 10 % with respect to control specimen were 6.59 %, 13.17 %, 21.56 % and 29.34 % respectively for OPC concrete beam specimens and for PPC concrete beam specimens 4.32 %, 10.65 %, 18.93 % and 26.04 % respectively.

From Figure 16 and Figure 17 it is exhibited that bond stress approximately drops for about 2.6 % and 2.1 % for (initial strain value) OPC and PPC concrete beam specimen and also 2 % and 2.1 % (final strain value) for OPC and PPC concrete beam specimens respectively for every percentage increase in corrosion level.

Table 6. *Bar force, reduced diameter and bond stress performance for different degree of corrosion in NBS beam*

Corrosion level, %	Bar force and Bond Stress values at slip region in strain values							
	Initial strain values				Final strain values			
	Micro Strain	Stress in bar, N/mm^2	Reduced Diameter, mm	Bond Stress, N/mm^2	Micro Strain	Stress in bar, N/mm^2	Reduced Diameter, mm	Bond Stress, N/mm^2
0	700	199.21	25.00	1.67	1330	367.96	25.00	3.08
2.5	660	188.85	24.69	1.56	1275	353.58	24.69	2.92
5	600	177.59	24.37	1.45	1210	337.08	24.37	2.75
7.5	560	163.25	24.04	1.31	1110	310.00	24.04	2.49
10	505	148.13	23.72	1.18	1050	293.93	23.72	2.33

Table 7. *Bar force, reduced diameter and bond stress performance for different levels of corrosion in PPC concrete beam specimens*

Corrosion level, %	Bar force and Bond Stress values at slip region (PPC concrete beam specimens)							
	Initial strain values				Final strain values			
	Micro Strain	Stress in bar, N/mm^2	Reduced Diameter, mm	Bond Stress, N/mm^2	Micro Strain	Stress in bar, N/mm^2	Reduced Diameter, mm	Bond Stress, N/mm^2
0	710	201.75	25.00	1.69	1400	387.00	25.00	3.24
2.5	690	196.46	24.69	1.62	1345	371.88	24.69	3.07
5	645	185.04	24.37	1.51	1280	354.85	24.37	2.89
7.5	585	170.13	24.04	1.37	1165	325.13	24.04	2.62
10	530	154.94	23.72	1.23	1085	303.13	23.72	2.41

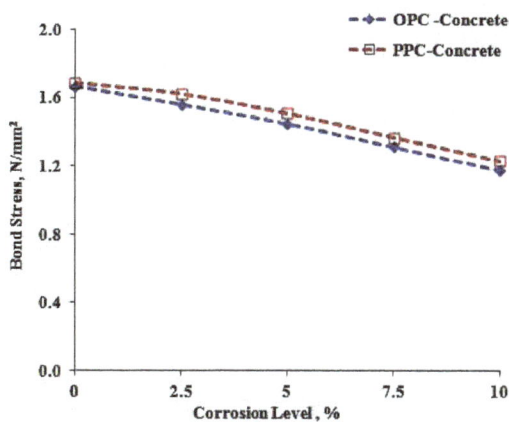

Figure 16. *Effect of corrosion levels on bond stress of slip region at final strain values*

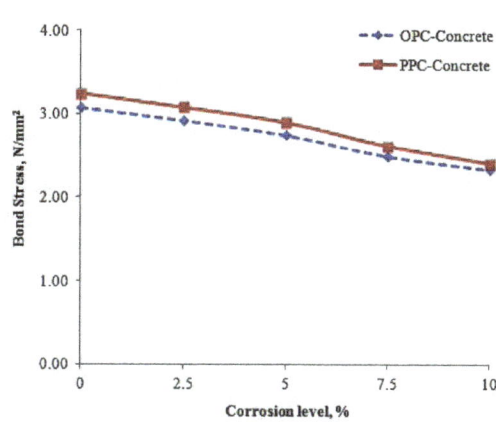

Figure 17. *Effect of corrosion levels on bond stress of slip region at final strain values*

Bond stress values for different degree of corrosion can be calculated from following equations obtained from Figure 16 and Figure 17, where *x* = corrosion levels, % and *y* = bond stress, N/mm^2. At initial slip point

(OPC concrete) $y = -0.049x + 1.678$ $R^2 = 0.996$ (9a)

(PPC concrete) $y = -0.046x + 1.718$ $R^2 = 0.982$ (9b)

At end slip point

(OPC concrete) $y = -0.076x + 3.099$ $R^2 = 0.992$ (10a)

(PPC concrete) $y = -0.084x + 3.269$ $R^2 = 0.990$ (10b)

Conclusions

Based on the detailed experimental investigations, following conclusions are drawn:

Reinforcement corrosion leads to the decline of load carrying capacity of NBS RC beam specimens. For every percentage increase in corrosion level, there is about 1.6 % decrease in load carrying capacity. For increasing corrosion level, strain values increase in the initial stages. Then at higher corrosion levels rate of increase of strain is higher for the same increment of load level, compared to the lower levels of corrosion. Crack width is less at the lower level of corrosion (0 to 5 %), after that it increases rapidly (7.5 to 10 %).

Reinforcement corrosion causes degradation of bond behavior. The strain value becomes large due to corrosion and the larger the corrosion lesser the bond stress value. Proposed regression equation is very much useful for quick assessment to predict the bond strength values for different corrosion levels in structure. Though the state of the art states about blended cements the research results done by OPC concrete is very much useful for the already existing structures subjected to corrosion.

References

[1]   S. Bhaskar, B. H. Bharatkumar, G. Ravindra, M. Neelamegam, *Journal of Structural Engineering (CSIR-SERC),* **37** (2010) 37-42.

[2]   T. Maheswaran, J. G. Sajavan, *Journal of Material Concrete Research*, **56** (2004) 359-366.

[3]   C. Andrade, C. Alonso, *Journal of Construction Building Materials*, **10** (1996) 315-328.

[4]   R. Park, T. Paulay, *Reinforced Concrete Structures*, John Wiley & Sons, Inc. New York (1975).

[5]   L. A. Lutz, P. Gergely, *ACI Materials Journal,* **11** (1967) 711-721.

[6]   IS 8112: 43 Grade Ordinary Portland cement – Specification Bureau of Indian Standards (1989).

[7]   IS 4031: Methods of physical tests for hydraulic cement, Bureau of Indian Standards (1988).

[8]   IS 383: Indian Standard Specification for Coarse and Fine aggregates from natural sources for Concrete (1970).

[9]   IS 456: Indian standards Code of Practice for Plain and Reinforced Concrete (2000).

[10] IS 10262: Recommended guidelines for concrete mix design, Bureau of Indian Standards (2009).

[11] R. J. Paul, Master Thesis, Department of Civil Engineering and Applied Mechanics, Mc Gill University Canada (1978) p. 66.

[12] S. Care, A. Raharinaivo, *Cement and Concrete Research*, **37** (2007) 1598–1612.

[13] S. Ahamad, *The Arabian Journal of Science and Engineering*, **34** (2009) 95-104.

[14] B. Pradhan, B. Bhattacharjee, *Construction and Building Materials,* **23** (2009) 2346- 2356.

[15] M. G. Fontana, *Corrosion Engineering*, Tata McGraw-Hill Education (2005).

Effect of electropolishing on vacuum furnace design

Sutanwi Lahiri[✉], Girish Kumar Sahu, Biswaranjan Dikshit, Radhelal Bhardwaj, Ashwini Dixit, Ranjna Kalra, Kiran Thakur, Kamalesh Dasgupta, Asoka Kumar Das and Lalit Mohan Gantayet*

Laser and Plasma Technology Division, Bhabha Atomic Research Centre, Mumbai-400085, India
*Beam Technology Development Group, Bhabha Atomic Research Centre, Mumbai-400085, India
[✉]Corresponding Author

Abstract

The use of thermal shields of materials having low emissivity in vacuum furnaces is well-known. However, the surface condition of the heat shields is one of the most important factors governing their efficiency as radiation resistances. The emissivity of the thermal shields dictates the power rating of the heaters in furnace design. The unpolished materials used in the heater tests showed poor performance leading to loss of a significant percentage of the input power. The present work deals with the refurbishment of the radiation heat shields used in a furnace for heating graphite structure. The effect of refurbishment of the heat shields by the buffing and subsequently electropolishing was found to improve the performance of the shields as heat reflectors. The composition of the electrolyte was chosen in such a way that the large shields of Mo, Inconel and SS can be polished using the same reagents in different ratios. The present work deals with the development of a standard electropolishing procedure for large metallic sheets and subsequently qualifying them by roughness and emissivity measurements. The improvement noted in the shielding efficiency of the furnace in the subsequent runs is also discussed here.

Keywords
Thermal shields; Etching; Emissivity

Introduction

The dominant mode of heat transfer in vacuum furnaces is radiation. The use of radiation heat shields to reduce the power loss from the hearth of a furnace is well known [1-4]. A set of heat shields made of various materials and assembled with minimum inter-shield separation, increases the net thermal resistance of the set. The most important factor governing the performance of each heat shield in the pack is the emissivity of both the surfaces, which, in turn, depends on their surface condition [1].

The vacuum furnace used in this study, has a pack of nine heat-shields as shown in figure 1. The pack of shields is designed to: (a) support heaters and (b) enclose the graphite hearth except a few view-ports. The innermost shield facing the hearth is of molybdenum and the next three shields were inconel-600 and the rest SS-304L. The Mo heater wires are laid on the innermost shield facing the graphite structure. The power input to the heaters is used to maintain the graphite structure at a temperature of 1500 K. The heat shields help in maintaining the hot zone at 1500 K without increasing the temperature of the chamber wall beyond 400 K. The vacuum chamber is maintained at a vacuum of the order of 10^{-5} mbar.

The primary objective of the experiment was to maintain the graphite hearth at a temperature of 1500 K. However, experiments with the unpolished heat shields as received from the fabricator failed to achieve the target temperature. Heat was being lost from the furnace primarily due to the high emissivity of the shields. The graphite structure could reach a temperature of 1200 K at 124 kW. On extrapolation, the required power to achieve 1500K was 320kW.

On further analysis, the presence of MoO and MoC was detected on the surface of the innermost shield. Carbon was present on all surfaces of the heat shields. The oxidation and the carburization is a result of the outgassed species from the graphite hearth. The data from the residual gas analyser (RGA) showed the presence of methane and carbon in significant quantities during the experiment. The presence of carbon and the formation of carbide resulted in an increase in their emissivity and therefore, loss of input power. The unpolished surface provides a larger area for carbon deposition and carbide formation and hence acts as a catalyst in the deterioration of its emissivity.

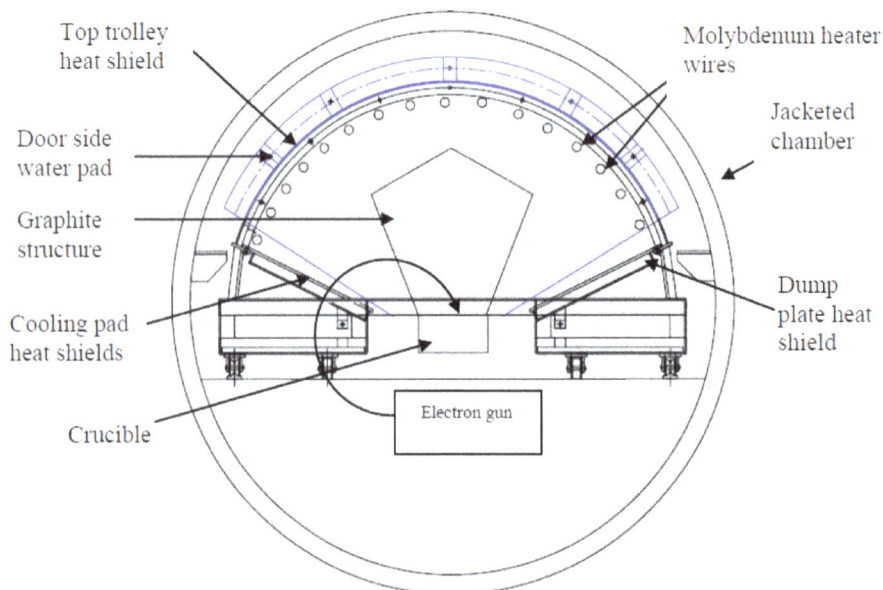

Figure 1. *Schematic diagram of the furnace*

The effective emissivity or reflectivity is a strong function of the surface finish [12-15]. As the temperature of the shield increases, the surface roughness needs to be significantly less than the peak wavelength of the heat radiation at working temperature to obtain high values of reflectivity. This calls for a highly polished surface.

Electropolishing is essentially the process of anodically smoothening of a surface in a suitable electrolyte. It removes the microscopic roughness on the surface, eliminating imperfections which trap and contain contaminants, in a more controlled manner compared to the other forms of chemical cleaning. Hence, electropolishing was chosen as the method of refurbishment of these shields. In short, the object to be electropolished is immersed in an electrolyte of an optimized composition and subjected to direct current [8].

Experimental

A flowchart summarizing the experimental procedure is given in Figure 2. It may be noted here that the range of average roughness values that electropolishing can achieve is 0.8 to 0.1 microns [18]. Therefore, 0.2 microns was selected for the roughness check.

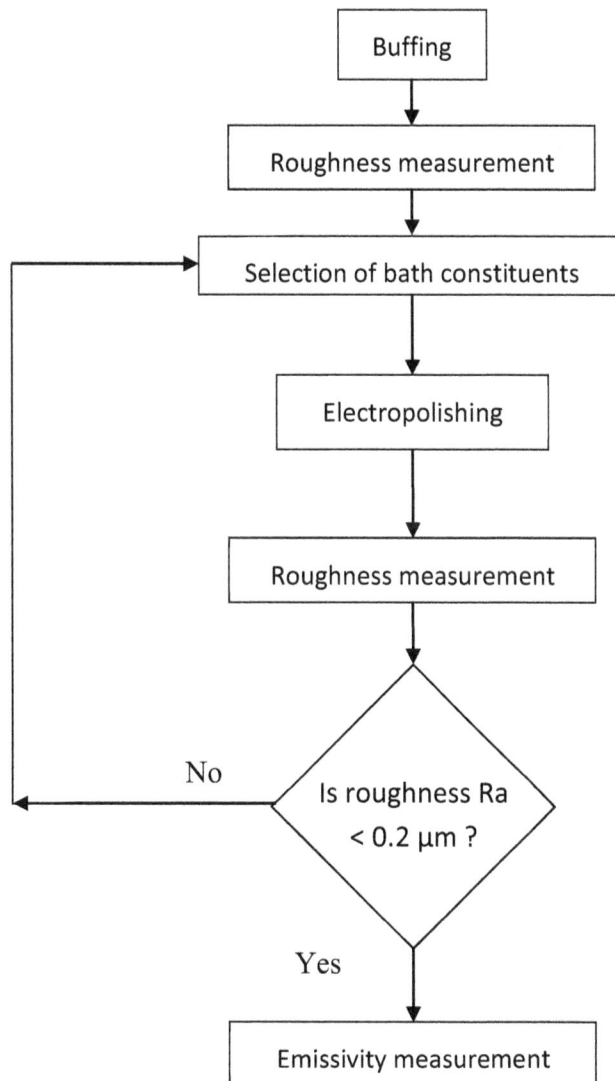

Figure 2: Flowchart of the experimental procedure

Selection of the bath constituents

There is a wide spectrum of electrolytes used in industrial application for electropolishing. The stainless steel has established electropolishing procedures with sulfuric acid and phosphoric acid [17]. However, some of the electrolytes reported for molybdenum and inconel (like potassium dichromate, *etc*) are too hazardous to be handled on a large scale [6-7]. Moreover, it is convenient to use the same bath constituents for polishing all the radiation heat shields by changing the ratio of the acids, instead of using separate baths and chemicals for separate materials. Since the process was to be carried out on large shields of 2000×500 mm area, a few trials were carried out on small samples to determine the optimum electropolishing conditions using the industrial solvents like sulfuric acid and phosphoric acid.

Pretreatment

Poor base metal conditions can result in less than optimum electropolished finishes. These flaws are revealed by electropolishing. Hence suitable pretreatment is required. In view of this, the shields were buffed with buffing wheel of grades 60 and 120. After the mechanical polishing and removal of burrs and sharp edges, the samples were washed in running tap water. The micrographs of the samples were obtained. Its roughness was measured and recorded.

Electropolishing

An electrolytic bath of dimensions 1500×1500×1000 mm was filled with a given composition of electrolyte. The copper plates of dimensions 400×200×2 mm was used as cathode. The voltage was set at a given value and the surface of the specimen (200×80×1 mm) was inspected after an interval of 3 minutes using a magnifying glass. The distance between anode and cathode was 800 mm. The current, voltage and time were noted. Temperature of the bath was monitored and found to rise to a steady state maximum temperature of 45 °C. It may be noted that there was no external agitation in the bath. After the specimen was removed from the bath, it was put under running tap water and dried in air for half an hour.

Emissivity measurement

One of the objectives of heat shield cleaning is to reduce the emissivity so that the heat loss due to absorption in the heat shields is reduced. The samples of Mo, Inconel and SS 304 sheets used in the heater runs and those obtained after electropolishing were made into filaments of suitable size. The bare wires of chromel and alumel are spot-welded on the filament and used as a thermocouple for measuring temperature as shown in Figure 3. The transmissivity of the glass window is known. Hence the difference in the temperature readings of the thermocouple (T_{actual}) was compared to the temperature measured by the pyrometer ($T_{measured}$) and the emissivity value ε of the filament was determined at 0.65 μm wavelength. The following expression was used for estimating emissivity [4]:

$$\frac{1}{T_{measured}} - \frac{1}{T_{actual}} = \frac{\lambda \ln(\tau\varepsilon)}{1.4388}.$$

a   **b**

Figure 3. *(a) Filaments made out of the heat shields (b) Hot filament as viewed by the pyrometer*

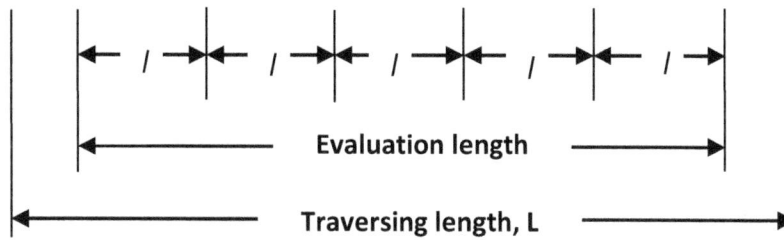

Figure 4. *Typical surface profile and evaluation length*

Roughness measurement

A token of area 50×50mm was made from each material and roughness was measured before and after electropolishing. The specimen is placed on a paper placed on a vibration less table. One edge of the specimen is aligned to the stand of a calibrated roughness meter and the stylus was used to scan the surface over a tracing length of 17.5 mm. The tracing length consists of pre-travel L_v, post travel L_n, and the evaluation length L_m. The evaluation length (in this case, 12.5 mm) consists of n (n = 5) number of sampling lengths (2.5 mm). The surface profile was recorded and analysed. A typical surface profile is shown in Figure 4.

Results and Discussion

Condition before electropolishing

It is observed that the surface condition of the virgin heat shields brought from the mill were not good enough to offer low value of emissivity. The imperfections acted as contamination traps and led to the deterioration of emissivity as the power loss from the hot zone was increased. The micrographs of the unpolished heat shields are shown in Figure 5.

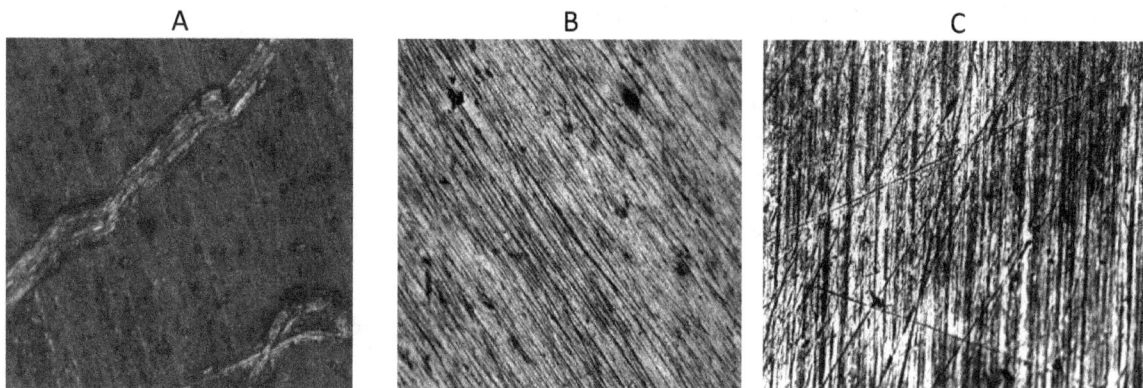

Figure 5. *Micrographs of the virgin heat-shields: A) Stainless steel B) Inconel C) Mo*

Electropolishing

The optimum electrolytic conditions for molybdenum, inconel and steel are given in Table 1. Several trials were carried out to determine the optimum conditions for electropolishing for each material. The ratio of the phosphoric acid and sulfuric acid was changed to determine the optimum composition for each material. It may be noted that the voltage across the electrodes was varied during the process of optimization. It was found that, the composition reported by Zamin *et al.* [16] was unsuitable for Mo sheets. The blue patches of molybdenum trioxide developed after electropolishing Mo was attributed to the oxidizing action of the sulfuric acid and hence the electropolishing was carried in orthophosphoric acid only.

Table 1. Optimum condition for electropolishing

Parameter	Molybdenum	Inconel	SS
Electrolyte	H_3PO_4 (85%)	H_3PO_4 (85%) and H_2SO_4 (5:1)	H_3PO_4 (85%) and H_2SO_4 (4:1)
Voltage, V	7	6	6
Time, min	6-10	5-8	5-8
Observation	Shining surface	Shining surface	Shining surface

Condition after electropolishing

It is observed that the micrographs of the electropolished heat shields show improved surface condition of all the three types of samples. Figure 6 presents a comparison of the micrographs showing the improvement caused by the exercise.

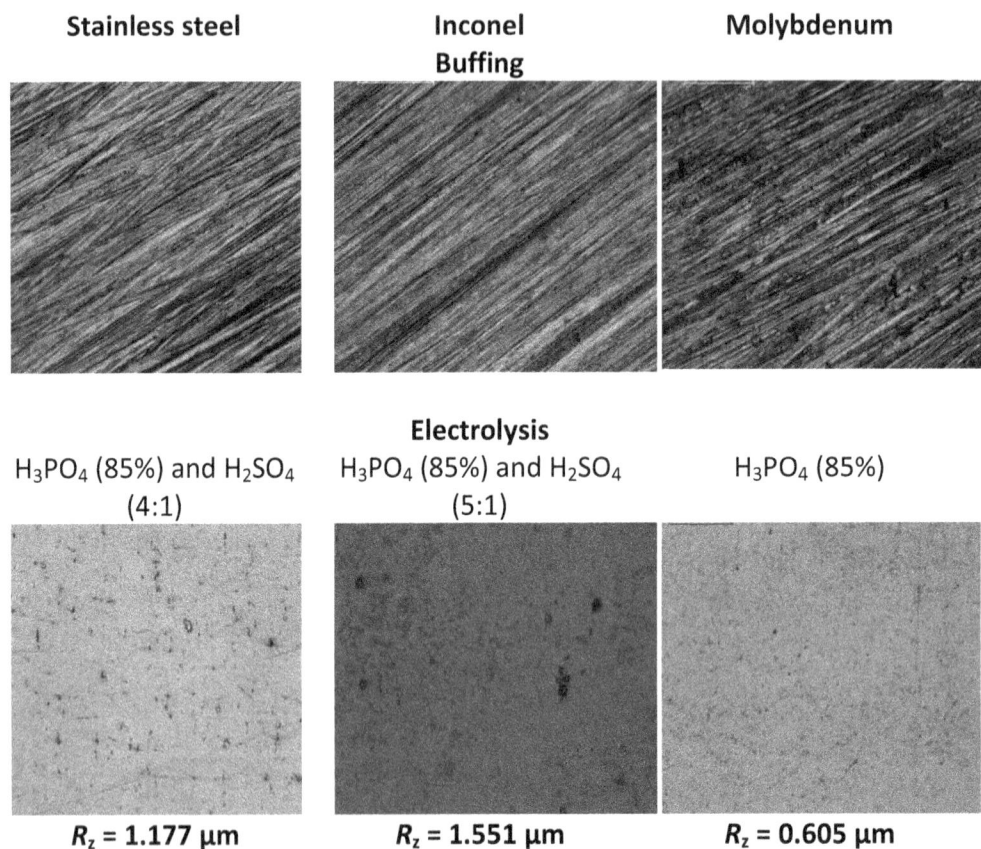

Figure 6. Micrographs of the electropolished heat-shields

Improvements in emissivity and roughness

As discussed earlier, emissivity of the surface is the most important parameter governing the performance of the shields. Since the roughness of the surface affects the emissivity values, an improvement of the surface profile was observed after the exercise. Table 2 gives the values of the surface roughness while Table 3 summarizes the improvement in emissivity caused by electro-polishing. The roughness values of the buffed samples before and after electropolishing were compared to determine the etch rates. The etch rate of molybdenum was found to be the fastest and that of SS was the slowest in their respective electrolytic baths.

This encouraged us to carry out similar polishing on the large shields used in the industrial furnace. The dimensions of the large heat shields vary from 500×500×1 mm to 2000×2000×1 mm.

Figure 7 shows the refurbished heat shields. The heat-shields are cleaned and polished to a roughness less than 2 μm. The effect of electropolishing was demonstrated in the performance of the heat shield in the subsequent heater test. The graphite structure was found to achieve the desired temperature at a much lower power compared to the previous test due to the improvement in the emissivity values (Table 4). Hence the objective of the refurbishment of the heat shields was fulfilled.

Table 2. Roughness values of the samples

Specimen	Roughness before electro-polishing $R_{z2}{'}$ μm	Roughness after electro-polishing R_{z2} / μm	Reduction in roughness R_{z1} / R_{z2}	Etch rate, (mg/cm^2) / min
SS-316	10.410	1.177	8.84	1.054
Inconel	8.162	1.551	5.26	1.116
Mo	9.380	0.605	15.5	1.800

Table 3. Emissivity values of the samples before and after refurbishment

Material	Emissivity at 0.65 μm	
	Before	After
SS-304	0.82	0.31
Inconel	0.78	0.5
Mo	0.7	0.27

Table 4. Heater power requirement

Parameter	Heat shield before refurbishment		Heat shield after refurbishment	
Heater power, kW	124	320	124	210
Temperature of graphite, K	1200	1523	1346	1546

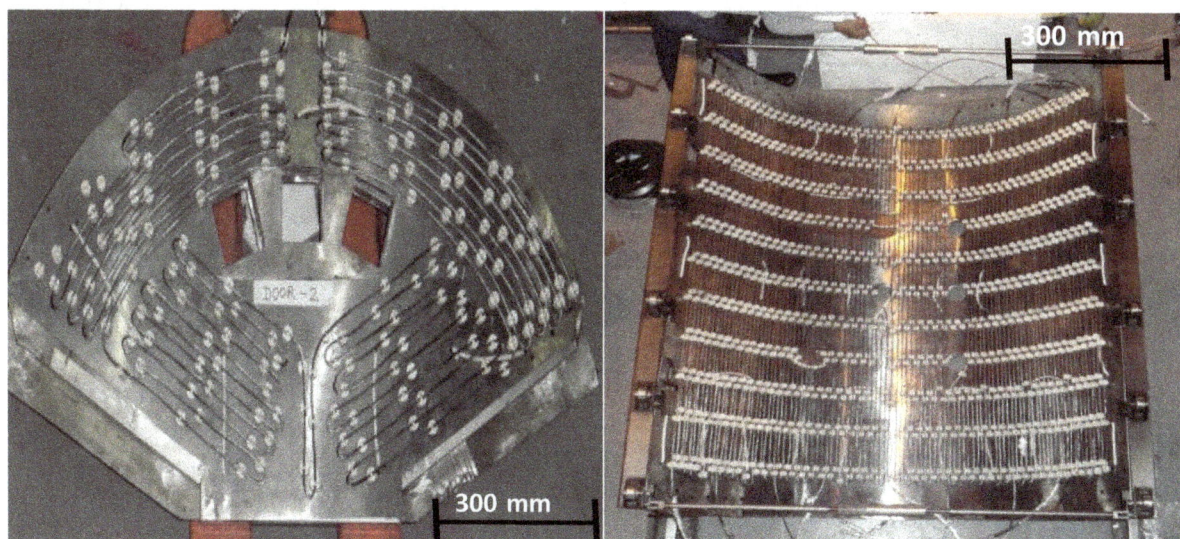

Figure 7. Electropolished heat shields

Conclusion

The present study gives a simple route for electropolishing of Mo and Inconel on a large scale using the commonly available laboratory chemicals. The results obtained in the trials were further analyzed and the roughness and emissivity of the surface was measured to qualify the polish. From Table 2, it is evident that the reduction in roughness has been the highest for Mo, followed by SS and Inconel. A good improvement of emissivity of the shields was observed. Since surface roughness, among other factors, influences the emissivity of a material and catalysis the surface reactions, a reduction in power fed to the heaters.

Electropolishing of heat shields reduces the heater power requirement of the furnace and therefore plays a significant role in the design and economics of vacuum furnaces. The power requirement came down by 33 % from the earlier 320 to 210 kW.

Acknowledgments: *The authors would like to thank Dr D. Das, Chemistry Division for his advice. The authors also thank Dr V. K. Mago for his valuable suggestions.*

References

[1] W. Espe, *Materials of High Vacuum Technology, Vol. 1: Metals and Metalloids*, Pergamon Press, New York, USA, 1966, p.360.

[2] F. P. Incropera,, D. P. DeWitt, T. L. Bergman, A. S. Lavine, *Fundamentals of Heat and Mass Transfer*, John Wiley & Sons, Hoboken, New Jersey, USA, 2007, p. 245.

[3] J. P. Holman, *Heat Transfer*, McGraw-Hill, New York, USA, 2009, p. 256.

[4] J. R. Howell, R. Siegel, M. P. Menguc, *Thermal Radiation Heat Transfer*, CRC Press, New York, USA, 2010, p. 250.

[5] C. Afonso, J. Matos, *Int. J. Refrig.* **29** (2006) 1144-1151.

[6] C. L. Faust, *Metal Finish.* **80** (1982) 21–25.

[7] K. B. Hensel, *Metal Finish.* **87** (1989) 89-96.

[8] D. A. Jones, *Principles and Prevention of Corrosion*, Macmillan Publishing Company, New York, NY, 1992, p 35-39.

[9] J. Mendez, R Akolkar, T. Andryushchenko, U. Landau, *J. Electrochem. Soc.*, **155(1)** (2008) D27-D34.

[10] Bing Du, Ian IvarSuni, *J. Electrochem. Soc.*, **151(6)** (2004) C375-C378.

[11] R. P. Allen, H. W. Arrowsmith, W. C. Budke, *Proc. 70th AIChE Annual Meeting,* New York, USA, November 13-17, 1977, Paper 109C.

[12] C. Wen, I. Mudawar, *Int. J. Heat Mass Tran.* **48** (2005) 1316-1329.

[13] C. Wen, I. Mudawar, *Int. J. Heat Mass Tran.* **49** (2006) 4279-4289.

[14] C. Wen, I. Mudawar, *Int. J. Heat Mass Tran.* **47** (2004) 3591-3605.

[15] S. Agababov, T. Vysokikh, *Temperature* **8** (1970) 770-773.

[16] M. Zamin, P Mayer, M K Murthy, *J. Electrochem. Soc.* **124(10)** (1977) 1557-1562.

[17] S. Habibzadeh, L. Li, D. Shum-Tim, E. C. Davis, S. Omanovic, *Corros. Sci.* **87** (2014) 89-100.

[18] E. P. Degamo, J. T. Black, R. A. Kohser, *Materials and Processes in Manufacturing,* John Wiley & Sons, Hoboken, New Jersey, USA, 2003, p.430.

13

Voltammetric studies on mercury behavior in different aqueous solutions for further development of a warning system designed for environmental monitoring

Paul-Cristinel Verestiuc, Igor Cretescu*,✉, Oana-Maria Tucaliuc,
Iuliana-Gabriela Breaban and Gheorghe Nemtoi**

Faculty of Geography and Geology, Al. I. Cuza University of Iasi, 20 A. Carol I Bd., Iasi, 700505, Romania
*Faculty of Chemical Engineering and Environmental Protection, Gheorghe Asachi Technical University of Iasi, 73, D. Mangeron Street, Iasi, 700050, Romania
**Faculty of Chemistry, Al. I. Cuza University of Iasi, 11, Carol I Bd., Iasi, 700506, Romania

✉Corresponding author

Abstract

This article presents some results concerning the electrochemical detection of mercury in different aqueous solutions, using the following electrodes: platinum-disk electrode (PDE), carbon paste electrode (CPE) and glass carbon electrode (GCE). Using the voltammetric technique applied on the above mentioned electrodes, the experimental conditions were established in order to obtain the maximum current peaks, in terms of the best analytical characteristics for mercury analyses. The dependence equations of cathodic current intensity on the scan rate were established in the case of mercury ion discharge in each prepared solution of 0.984 mM $HgCl_2$ in different electrolyte background: 0.1 M KCl, 0.1 M H_2SO_4 and 0.9 % NaCl. Among the three investigated electrodes, the carbon paste electrode presented the highest detection sensitivity toward mercury ions in the aqueous solution. It was observed that, at a low scanning rate, the pH had an insignificant influence over the current peak intensity; however, the quantification of this influence was achieved using a quadratic polynomial equation, which could prevent the errors in mercury detection in case of industrial waste stream pH changes. The calibration curves for mercury in 0.9 % NaCl solution and in the tap water respectively were carried out.

Keywords
Cathodic linear voltammetry; platinum electrode; carbon paste electrode; glass carbon electrode; mercuric ion; flow electrochemical cell.

Introduction

The ubiquitous presence of mercury in the environment is due to both natural geological activities as well as due to increasing anthropogenic pollution. Because of its unique electronic configuration, mercury behaves similarly to noble gas elements, but the physical and chemical properties of mercury such as high surface tension, high specific gravity, low electrical resistance, and a constant volume of expansion over the entire temperature range in liquid state can rapidly transform this element into a hazardous air pollutant [1,2].

The Water Framework Directive (2000/60/EC) classified mercury as a priority hazardous substance, establishing that from 2015, no more mercury from production processes can be discharged [3, 4].

Mercury exists in a large number of forms, *i.e.* as "elemental" Hg(0), monovalent or divalent mercury Hg(I) and Hg(II), and in inorganic and organic compounds. Metallic mercury Hg(0) and most mercury compounds present high toxicity, acting as a bioaccumulative neurotoxin [5,6].

Mercury ions are strongly adsorbed by soils or sediments in acid medium and are slowly desorbed, due to the content of clay minerals and/or organic matter, which are responsible for its behavior. The reaction products resulting from the methylation of inorganic mercury forms impose a significant risk to humans and wildlife due to tendency to accumulate in the food chain, and their ability to act as neurotoxins [6,7]. Exposure to various forms of mercury will harm human health. Moderate and repeated exposure to organic forms (lower than a few mg m^{-3} Hg, but higher than 0.05 mg m^{-3} Hg) causes symptoms of poisoning such as: lack of coordination of movement, impairment of peripheral vision, speech, hearing or walking as well as muscle weakness. Inhaled or physical contact with inorganic forms cause: tremors, emotional or neuromuscular changes, insomnia, headaches, disturbances in sensation, changes in nerve responses, and performance deficits on tests of cognitive function. With prolonged or high concentration exposure, kidney effects, respiratory failure and death may occur [8,9].

Mercuric chloride (HgCl$_2$) is used as a depolarizer in electric batteries and as a reactant in organic synthesis and analytical chemistry [10]. The presence of this element in different environmental components could be considered as harmful to human health well environmentally dangerous due to the mercury content.

Taking into consideration all the above mentioned aspects, the detection of mercuric ion has become a priority for environmental safety and human health. The determination of trace amounts of mercury, has led to some analytical problems because it can be found in several chemical forms [11]. For an accurate determination of mercury at trace and ultra-trace levels, analytical methods with high sensitivity and selectivity are needed. There are a number of analytical methods for mercury detection which require expensive instruments, well-controlled experimental conditions, sample preparation and relatively large sample volumes [12]. Electrochemical detection of trace metals offers important advantages, such as remarkable sensitivity, inherent miniaturization and portability, remote monitoring and decentralized measurements, low cost and compatibility with turbid samples [13,14]. Therefore, electrochemical methods are less costly and require no sophisticated equipment.

Chemically modified electrodes have received increasing attention, which has led to improvements in the sensitivity and selectivity of electrochemical analysis techniques in the recent decades. The determination of Hg(II) ions using chemically modified electrodes has been investigated recently, using plants [14], gold film [15], polymer films [16] or organic compounds with chelating groups [17].

The behavior of electrodes has also been studied using modified a carbon paste electrode [18], modified glass carbon electrode [17] or modified gold nanoelectrode ensembles [19].

A relatively recent review of electrochemical sensors and detectors has been done by Bakker [20], characterizing this area of research as being one of the most fruitful and interdisciplinary areas of research in analytical chemistry. A recent paper by Pujol [21] analyzed the actual state of art concerning the sensors and devices for heavy metals detection in water, but the mercury detection from environmental samples was not studied in details.

For this reason in this study it was investigated the cathodic discharge of the mercuric ion on different type of electrodes such as: platinum-disk electrode, carbon paste electrode and glass carbon electrode, in respectively different aqueous solutions (0.1 M KCl, 0.1 M H_2SO_4, 0.9 % NaCl) by the voltammetric method in order to find the most suitable conditions for analytical purpose.

EXPERIMENTAL

In order to simulate the wastewater from the industrial stream, aqueous solutions of $HgCl_2$ in 0.1 M H_2SO_4 at a pH value of 0.99, 0.9 % NaCl at a pH value of 5.66 and 0.1 M KCl at a pH value of 7.43 was prepared, using analytical purity reagents and double distilled water. The investigation of mercury behavior in these solutions was carried out using voltammetric measurements [22-26] with the above mentioned electrodes (GCE, CPE and PDE) individually. These electrodes play the role of working electrode (**WE**) in a flow electrochemical cell (as is presented in Fig. 1) through which the samples from the industrial stream are passed using a peristaltic pump. A saturated calomel electrode (**SCE**) was used as the reference electrode (**RE**), while a platinum electrode was used as the counter (auxiliary) electrode (**CE**) at the temperature of 25 °C. Also, pH and temperature sensors were located in the electrochemical cell in order to provide corrections in the case of possible changes in the above mentioned parameters which should be kept constant at the same values as were used during the calibration procedure.

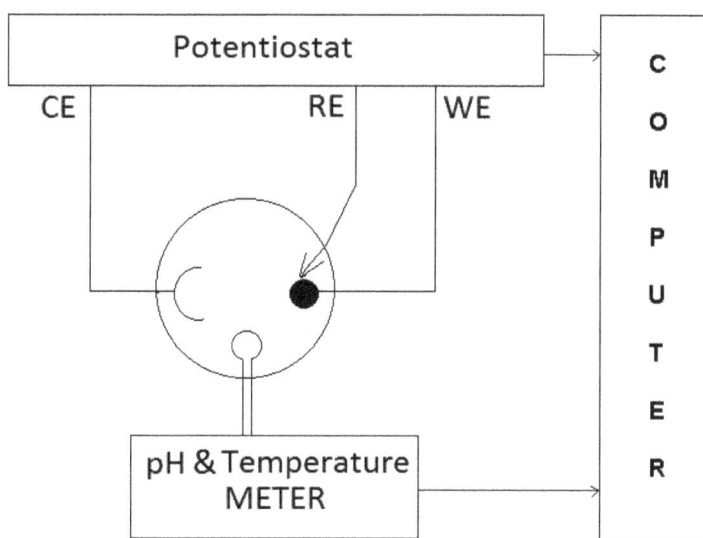

Figure 1. Experimental setup for study of the voltammetric detection of mercury in aqueous solution

The pH corrections were achieved using software (the pH dependence of the current peak was determined by a simple equation), while the temperature corrections were first achieved by the thermostatic system for sample processing and finally by the software corrections for the fine pH adjustments.

The voltammograms were recorded using the potentiostat VoltaLab 32 (Radiometer Analytical) [27-29] after stopping the flow of liquid samples through the electrochemical cell (using a solenoid mini-valve) and nitrogen bubbling in the investigated solutions, in order to remove the dissolved oxygen.

The investigated electrodes were used as purchased from Radiometer Analytical, except for the carbon paste electrode, which was prepared according to methods previously presented in the literature [30-31]. The electrode surfaces were: $3.14 \cdot 10^{-2}$ cm^2 for PDE; $7.07 \ 10^{-2}$ cm^2 for GCE and $19.63 \ 10^{-2}$ cm^2 for CPE.

Results and discussion

In Figures 2, 3 and 4, the cathodic linear voltammograms are presented, corresponding to the reduction of Hg^{2+} on the PDE (Φ = 2mm) at different scanning rates in different aqueous media.

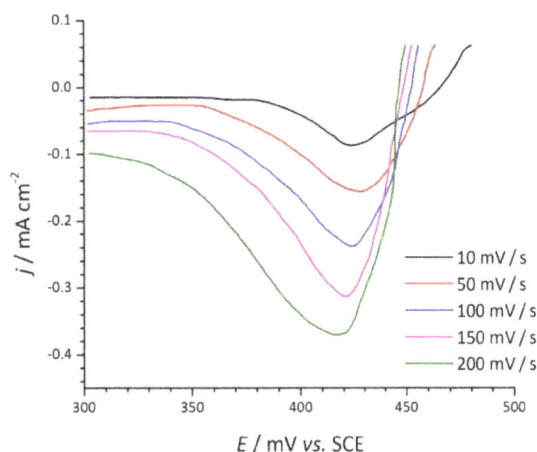

Figure 2. Linear cathodic voltammograms on PDE in 0.1 M H$_2$SO$_4$ aqueous solution recorded at different scanning rates

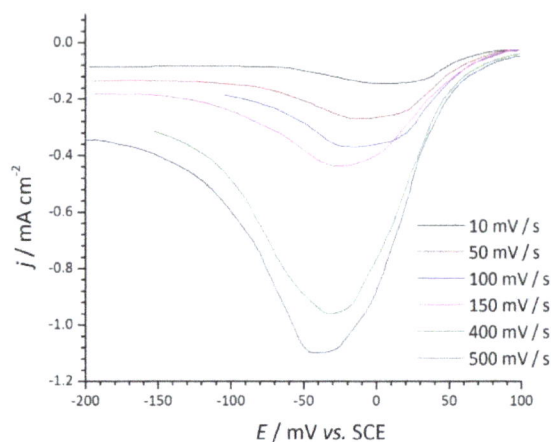

Figure 3. Linear cathodic voltammograms on PDE in 0.1 M KCl aqueous solution recorded at different scanning rates

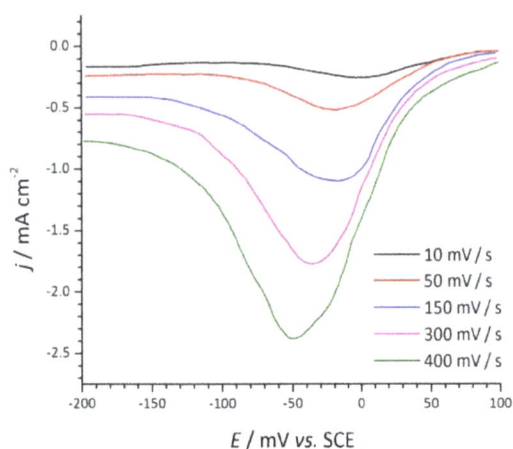

Figure 4. Linear cathodic voltammograms on PDE in 0.9 % NaCl aqueous solution recorded at different scanning rates.

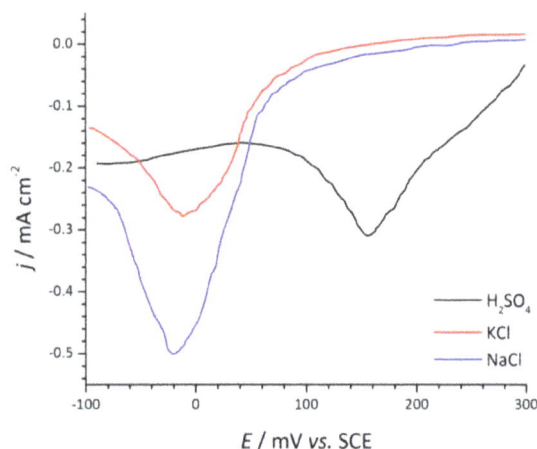

Figure 5. Linear cathodic voltammograms of mercury ion discharge on PDE in different solutions at constant scanning rate of 50 mV s^{-1}

Concerning the electrochemical behavior of mercuric ions in the three different aqueous solutions shown in Figures 2-4, it was observed that there was a simultaneous increase in PDE sensitivity with the scanning rate, at a concentration of 0.984 mM, as was expected.

This behavior leads to the following dependence equations for the cathodic peaks, which express the discharge of the mercuric ion on the platinum electrode, in all investigated solutions:

a. 0.1 M H_2SO_4: $I = -2.114 - 0.058\,v + 3.474 \times 10^{-5}\,v^2$ ($R = 0.99952$)   (1)

b. 0.1 M KCl: $I = -4.285 - 0.072\,v + 2.168 \times 10^{-5}\,v^2$ ($R = 0.99757$)   (2)

c. physiological serum (0.9 % NaCl): $I = -6.815 - 0.140v - 6.611 \times 10^{-5}\,v^2$ ($R = 0.99827$)   (3)

where: I = intensity current (mA), v = scanning rate (mV s^{-1}), R = correlation coefficient

According to the experimental results, it was noted that the highest sensitivity of PDE was obtained in physiological serum, followed by the solution prepared based on sulfuric acid and potassium chloride, respectively (Figure 5).

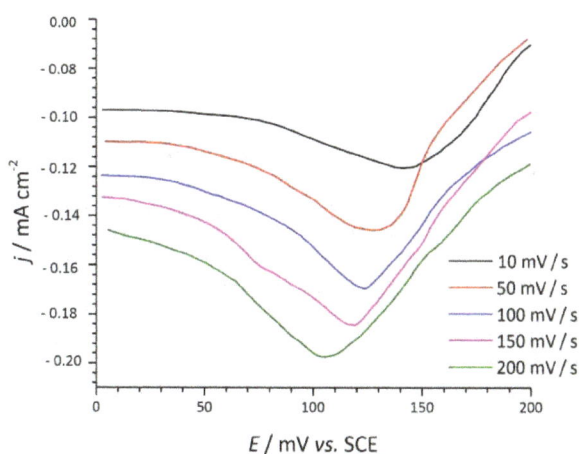

Figure 6. Linear cathodic voltammograms on CGE in 0.1 M H_2SO_4 aqueous solution recorded at different scanning rates.

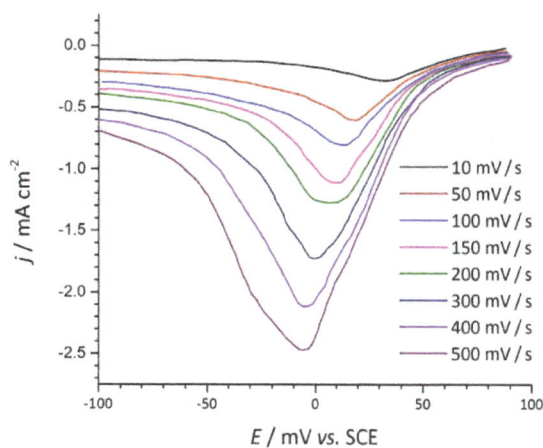

Figure 7. Linear cathodic voltammograms on CGE in 0.1 M KCl aqueous solution recorded at different scanning rates

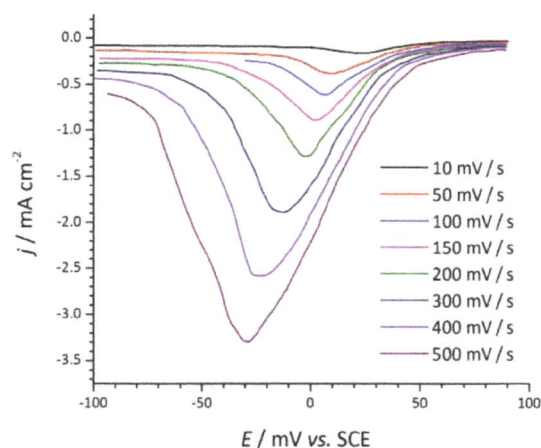

Figure 8. Linear cathodic voltammograms on CGE in 0.9 % NaCl aqueous solution recorded at different scanning rates

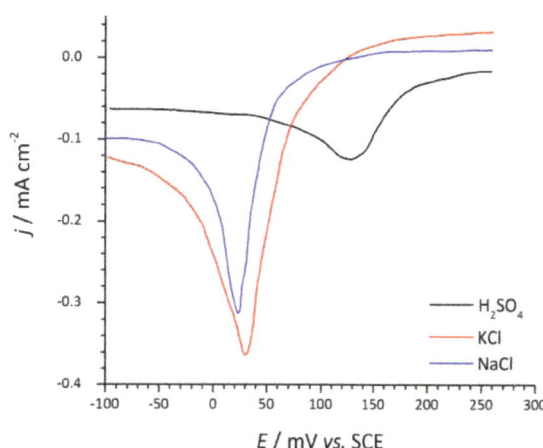

Figure 9. Linear cathodic voltammograms of mercury ion discharge on GCE in different aqueous solution at constant scanning rate of 50 mV/s

Figures 6-8 present the cathodic linear voltammograms for each of the aqueous solutions recorded on the GCE at different scanning rates.

In this case, the following dependence equations for the cathodic peak were established, which express the discharge of the mercuric ion (0.984 mM) on the platinum disc electrode, in all three investigated solutions:

a. 0.1 M H_2SO_4: $I = -5.7573 - 0.06488 + 1.3139 \times 10^{-4} v^2$ ($R = 0.99404$)   (4)

b. 0.1 M KCl: $I = -11.895 - 0.239 v + 1.079 \times 10^{-4} v^2$ ($R = 0.998$)   (5)

c. physiological serum (0.9 % NaCl): $I = -3.745 - 0.347 v + 1.355 \times 10^{-4} v^2$ ($R = 0.99918$)   (6)

where: I = current intensity (mA), v = scanning rate (mV s^{-1}), R = correlation coefficient

In accordance with the Figure 9, the sensitivity of the GCE toward the mercuric ion was highest in KCl aqueous solution and lowest in sulfuric acid.

The entire cathodic process of mercuric ion discharge on the CPE in all of the three investigated solutions is presented in detail in linear voltammograms (Figures 10-12).

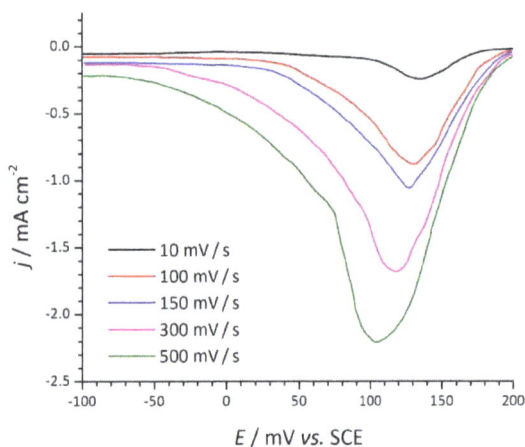

Figure 10. Linear cathodic voltammograms on CPE in 0.1 M H_2SO_4 aqueous solution recorded at different scanning rates.

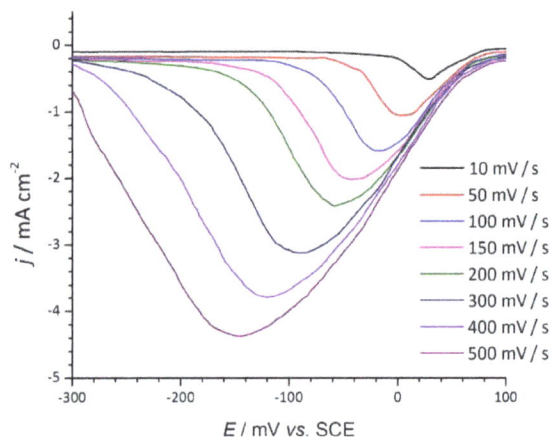

Figure 11. Linear cathodic voltammograms on CPE in 0.1 M KCl aqueous solution recorded at different scanning rates

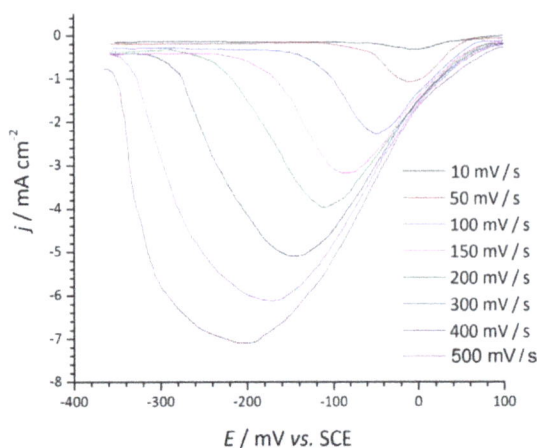

Figure 12. Linear cathodic voltammograms on CPE in 0.9 % NaCl aqueous solution recorded at different scanning rates

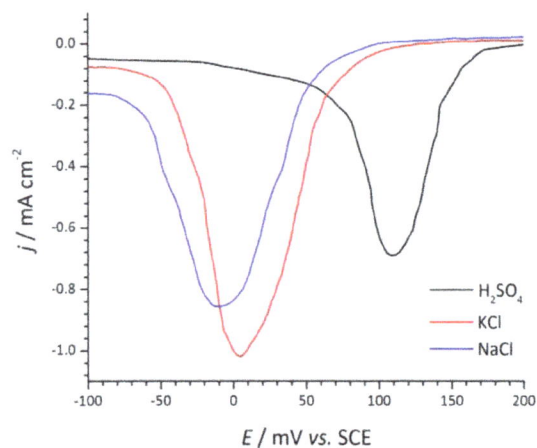

Figure 13. Linear cathodic voltammograms of mercury ion discharge on CPE in different solutions at constant scanning rate of 50mV s^{-1}

Data obtained using the CPE revealed the same behavior as with the GCE regarding higher sensitivity, as follows: 0.1 M KCl, physiological serum and 0.1 M H_2SO_4 (Figure 13).

The measurements performed on the CPE, in each investigated solution led to the following equations:

a. 0.1M H_2SO_4: $I = -38.137 - 1.282\ v + 9.944\times10^{-4}\ v^2$ ($R = 0.99581$)   (7)

b. 0.1 M KCl: $I = -80.923 - 2.168\ v + 1.27\times10^{-3}\ v^2$ ($R = 0.99808$)   (8)

c. physiological serum (0.9 % NaCl): $I = -34.086 - 3.25\ v + 2.23\times10^{-3}\ v^2$ ($R=0.99867$)   (9)

where: I = current intensity (mA), v = scanning rate (mV s^{-1}), R = correlation coefficient

These equations point out the dependence of the current intensity (corresponding to the discharge of mercuric ions in aqueous solutions at a concentration of 0.984 mM $HgCl_2$) on scanning rates.

Increasing the scanning rate during investigations into mercuric ion discharge for each studied electrode in the aqueous solutions led to a change in the cathodic potential values, which were shifted to negative values, suggesting a quasi-reversible process.

Among the electrodes used in this study, the carbon paste electrode was the most sensitive with respect to the cathodic discharge of mercuric ions in all the investigated aqueous solutions. Taking into consideration this aspect, the pH dependence of the current peak (Figure 14) and the calibration curves (Figure 15) in the NaCl solution and tap water, respectively, were presented only for the CPE.

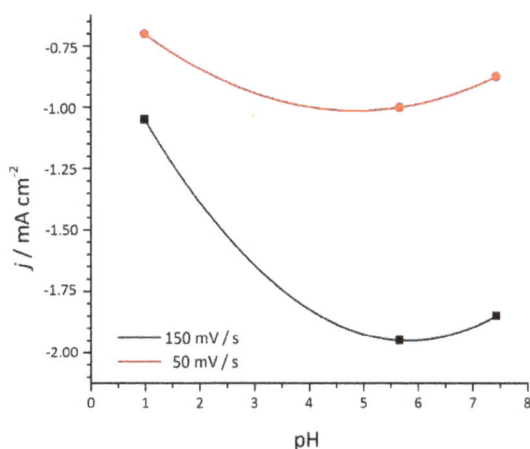

Figure 14. Influence of solution pH on the cathodic peak on CPE in 0.9 % NaCl aqueous solution recorded at two scanning rates

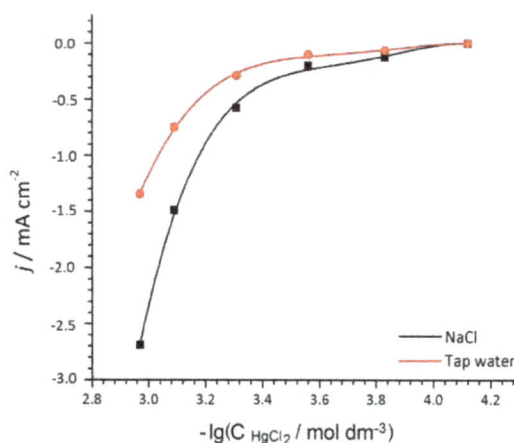

Figure 15. Calibration curves for determination of $HgCl_2$ using CPE in 0.9 % NaCl aqueous solution and tap water, respectively

As presented in Figure 14, a low pH dependence of the current peak was observed for a scanning rate of 50 mV s^{-1}, but if this value was multiplied by three fold, the dependence became more evident, especially in the acidic region. These equations are presented as follows:

$I_{50} = -0.12847 - 0.09001\ \text{pH} + 0.00774\ \text{pH}^2$ (10)

$I_{150} = -0.10381 - 0.0407\ \text{pH} + 0.00419\ \text{pH}^2$ (11)

The calibration curves presented in Figure 15 pointed out an approximately linear dependence for both calibration curves, which are separated into two regions. However, these calibration curves could be very well fitted following quadratic polynomial equations:

$X = -\lg[HgCl_2]$

$$I_{NaCl} = -242.414 + 257.620\ X - 102.678\ X^2 + 18.175\ X^3 - 1.205\ X^4 \qquad (12)$$

$$I_{tap\ water} = -121.207 + 128.825\ X - 51.339\ X^2 + 9.087\ X^3 - 0.602\ X^4 \qquad (13)$$

The decreased sensitivity of the voltammetric method for mercury detection in real samples (tap water provided from the Iasi drinking water treatment plant) was evident, but the analytical signal was sufficiently high to be discriminated from the global noise determined in such a complex matrix.

The response limit of the concentration peak for Hg (II) detection was in the range of 10^{-5} M to 10^{-3} M. The optimal operating conditions were established after a period of three minutes of preconcentration.

Conclusions

The following conclusions can be drawn for the voltammetric studies regarding the behavior of mercuric ions in the presence of different supporting electrolytes in aqueous solutions using different electrodes:

- The cathodic discharge of mercury ions from $HgCl_2$ aqueous solutions in different electrolyte support media such as 0.1 M KCl, physiological serum (0.9 % NaCl), and 0.1 M H_2SO_4, using different electrode materials (platinum, glass carbon and carbon paste) was evidenced by a relevant current peak, whose height (in terms of the analytical method sensitivity) is dependent on the potential scanning rate;
- The cathodic discharge of mercuric ions, in all investigated aqueous solutions, was a slow and quasi-reversible process, on all three of the investigated electrode materials;
- The best electrode was the carbon paste electrode, in terms of the sensitivity toward mercuric ions, in concordance with the equations describing the dependence between the cathodic peak intensity and the potential scanning rate. It was pointed out that the CPE presented the highest detection sensitivity toward mercuric ions in an aqueous solution, working in a wide pH range and having a short response time. Besides these aspects, the CPE is very easily prepared and has good reproducibility and repeatability for mercury analysis, which recommend it for this analytical application;
- Based on the results, a miniaturized voltammetric flow cell will be developed as an integrated component of the equipment used for mercury detection/monitoring in industrial waste streams. Based on the mercury detection capacity, a warning system will be designed to avoid mercury discharge into surface waters in the case of accidentally increasing the mercury concentration.
- Due to the increased capacity for mercury detection by this proposed monitoring system, based on a voltammetric method and an electrochemical flow cell, its versatility could be extended using some auxiliary equipment to other environmental components such as air and soil.
- At low scanning rate, the pH had an insignificant influence over the current peak intensity in the case of the CPE in 0.9 % NaCl solution.

- In order to avoid errors in mercury detection in cases of industrial stream pH changes, the quantification of the pH influence was achieved using a quadratic polynomial equation.
- The calibration curves for mercury in a 0.9 % NaCl solution and in tap water had similar shapes (which could be divided into two linear sections) but with decreased the sensitivity in tap water.

Acknowledgements: *This work was supported by the strategic grant POSDRU/159/1.5/S/133652, co-financed by the European Social Fund within the Sectorial Operational Program Human Resources Development 2007 – 2013.*

References

[1] R. P. Mason, W. F. Fitzgerald, F. M. M. Morel, *Geochimica et Cosmochimica Acta* **58 (15)** (1994) 3191-3198

[2] W. H. Schroeder, J. Munthe, *Atmospheric Environment* **32 (5)** (1998) 809-822

[3] A. Pichard, *Mercury and its derivates,* Institut National de L'environment Industiel et des Risques, Amiens, France, 2000, p. 6-8

[4] J. De Clerq, International Journal of Industry Chemistry **3 (1)** (2012)

[5] O. Lindquist, K. Johansson, M. Aastrup, A. Andersson, L. Bringmark, G. Hovsenius, L. Hakanson, A. Iverfeldt, M. Meili, B. Timm, *Water, Air, & Soil Pollution* **55 (1)** (1991) 23-32

[6] H. Satoh, *Industrial Health* **38 (1)** (2000) 153-164

[7] W. F. Fitzgerald, T.W. Clarkson, *Environmental Health Perspectives* **96 (1)** (1991) 159-166

[8] U. S. Environmental Protection Agency, *Mercury Study Report to Congress*, Volume III, 1997, p. 93

[9] T. W. Clarkson, L. Magos, G. J. Myers, *The New England Journal of Medicine* **349 (18)** (2003) 1731-1737

[10] W. T. S Zaugg, D. Foreman, L. M. Faires, M. G. Werner, T. J. Leiker; P. F. Rogerson, *Environmental Science & Technology* **26 (7)** (1992) 1307-1314

[11] D. W. Boeing, *Chemosphere* **40 (1)** (2000) 1335 – 1351

[12] O. M. Tucaliuc, I. Cretescu, G. Nemtoi , I. G. Breaban, G. Soreanu, O. G. Iancu, *Environmental Engineering and Management Journal* **13 (8)** (2014) 2051-2061

[13] J. Wang, B. Tian, J. Wang J. Lu, C. Olsen, C. Yarnitzky, K. Olsen, D. Hammerstrom, W. Bennet, *Analytica Chimica Acta* **385 (1)** (1999) 429-435

[14] D. S. Rajawat, S. Srivastava, S. P. Satsangee, *International Journal of Electrochemical Science* **7 (1)** (2012) 11456-11469

[15] R. D. Riso, M. Waeles, P. Monbet, C. J. Chaumery, *Analytica Chimica Acta* **410 (1)** (2000) 97-105

[16] L. R. Popescu, E. M. Ungureanu, G. O. Buica, C. Dinu, M. Iordache, *Scientific Bulletin-University Politehnica of Bucharest* Series B **75(4)** (2013) 1454-2331

[17] F. Wang, J. Liu, Y. J. Wu, Y. M. Gao, X.F. Huang, *Journal of the Chinese Chemical Society* **56 (4)** (2009) 778-784

[18] C. M. V. B. Almeida, B.F. Giannettio, *Electrochemistry Communications* **4 (1)** (2002) 985-988

[19] B. K. Jena, C. R. Raj, *Analytical Chemistry* **80 (13)** (2008) 4836-4844

[20] E. Bakker, M. Telting-Diaz, *Analytical Chemistry* **74 (1)** (2002) 2781-2800

[21] L. Pujol, D. Evrard, K. Groenen-Serrano, M. Freyssinier, A. Ruffien-Cizsak, P. Gross, *Frontiers in Chemistry* **2 (19)** (2014) 1-24

[22] Y. M. Sin, W. F. Teh, M. K. Wong, P. K. Reddy, *Bulletin of Environmental Contamination and Toxicology* **44 (1)** (1990) 616-622

[23]   D. M. Gordin, T. Gloriant, G. Nemtoi, R. Chelariu, N. Aelenei, A. Guillou, D. Ansel, *Materials letters* **59 (23)** *(*2005) 2936-2941

[24]   D. Mareci, C. Bocanu, G. Nemtoi, D. Aelenei, *Journal of the Serbian Chemical Society.* **70 (6)** (2005) 891-897

[25]   G. Nemtoi, A. Ciomaga, T. Lupascu, *Revue Roumaine de Chimie* **57 (9-10)** (2012) 837-841

[26]   A. V. Sandu, A. Ciomaga, G. Nemtoi, C. Bejenariu, I. Sandu, *Microscopy Research and Technique* **75(12)** (2012) 1711-1716

[27]   G. Nemtoi, H. Chiriac, O. Dragoş, M.O. Apostu, D. Lutic, *Acta Chemica Iasi* **17 (2)** (2009) 151-168

[28]   G. Nemtoi, M. S. Secula, I. Cretescu, S. Petrescu, *Revue Roumaine de Chimie* **58 (12)** (2007) 655-659

[29]   G. Nemtoi, D.I. Cuciurean, *Revista de Chimie* **53 (2)** (2002) 146-149

[30]   S. I. M. Zayed , H. A. M. Arida, *International Journal of Electrochemical Science* **8** (2013) 1340 – 1348

[31]   J. Zima, I. Svancara, J. Barek, K. Vytras, *Critical Reviews in Analytical Chemistry* **39** (2009) 204–227

14

Use of hydrous titanium dioxide as potential sorbent for the removal of manganese from water

Ramakrishnan Kamaraj, Pandian Ganesan and Subramanyan Vasudevan⊠

CSIR-Central Electrochemical Research Institute, Karaikudi - 630 006, India

⊠Corresponding Author

Abstract

This research article deals with an electrosynthesis of hydrous titanium dioxide by anodic dissolution of titanium sacrificial anodes and their application for the adsorption of manganese from aqueous solution. Titanium sheet was used as the sacrificial anode and galvanized iron sheet was used as the cathode. The optimization of different experimental parameters like initial ion concentration, current density, pH, temperature, etc., on the removal efficiency of manganese was carried out. The maximum removal efficiency of 97.55 % was achieved at a current density of 0.08 A dm^{-2} and pH of 7.0. The Langmuir, Freundlich and Redlich Peterson isotherm models were applied to describe the equilibrium isotherms and the isotherm constants were determined. The adsorption of manganese preferably followed the Langmuir adsorption isotherm. The adsorption kinetics was modelled by first- and second- order rate models and the adsorption kinetic studies showed that the adsorption of manganese was best described using the second-order kinetic model. Thermodynamic parameters indicate that the adsorption of manganese on hydrous titanium dioxide was feasible, spontaneous and exothermic.

Keywords:

Titanium dioxide; manganese; adsorption; thermodynamics; isotherm; kinetics.

Introduction

Manganese (Mn) is a naturally occurring element found in the air, soil, and water. It can exist in seven oxidation states ranging from -2 to +7. It is rarely found in its elemental state and is therefore a component of over 100 minerals and exists mainly as oxides, carbonates, and silicates. Its most common mineral is pyrolusite (MnO_2). In ground water, manganese is a common contaminant and its presence is due to leaching processes and varies widely depending on the rock type. Also, manganese has a variety of applications such as in ceramics, metallurgical processes, mining, dry cell batteries, pigments and paints which all can be the sources of underground pollution [1]. In addition to the disposal of untreated discharge from above the applications into water, another major source of pollution of the manganese is burning of coal and oil [2]. Manganese is an essential metal for the human system and many enzymes are activated by manganese. The manganese contaminant in ground water affects the intelligent quotient (IQ) of children. Intake of higher concentrations of manganese causes neuro toxic disease like Parkinsonism and manganese psychosis, an irreversible neurological disorder [3–5]. The prolonged over intake potentially affects the central nervous system, lungs, also causes diseases of disturbed speech called prognosis, also cause bronchitis and pneumonia [6,7]. The World Health Organization (WHO) prescribed the permissible limit for the manganese in the ground water is 0.05 mg L^{-1}. For this reason, there is great interest in the development of environmentally clean methods to destroy such compounds in aqueous medium for avoiding their dangerous accumulation in the aquatic environment.

Because of its high solubility over a wide pH range, toxicity and non-degradable nature, it is notoriously difficult to remove manganese from contaminated water [8,9]. Hence the researchers in the world have carried out significant work on their removal from aqueous solutions and industrial effluents [10-16]. The usual method for removing toxic metals from water include electrodialysis, chemical coagulation, reverse osmosis, co-precipitation, complexation, solvent extraction, ion exchange, electrochemical treatment and adsorption. Physical methods like ion exchange, reverse osmosis and electrodialysis have proven to be either too expensive or inefficient to remove manganese from water. At present, chemical treatments are not used due to disadvantages like high costs of maintenance, problems of sludge handling and its disposal, and neutralization of effluent. In this scenario, the electrochemical technologies have received great attention for the prevention of pollution problems, as reported in several reviews [17-19].

In the recent decade, electrodissolution process, where the coagulants generated *in-situ*, has been increasingly used in the world for treating the industrial wastewater, ground water and surface water and many studies conducted to optimize this process for specific problems [19]. The sacrificial anodic electrodes, commonly consisting of iron and aluminum, are used to continuously supply metallic ions as the source of coagulants, which can hydrolyze near the anode to form a series of metallic hydroxides capable of destabilizing dispersed particles. This process generates large quantities of iron and aluminum salt coagulated sludge, which inhibits efficient water treatment. From the generated coagulant, nothing may be recovered or reused, and require further incineration and landfill treatment. Furthermore, the appearance of dissolved iron in aquatic suspensions can lead to visual, odor and taste problems resulting from later growth of iron bacteria [20]. Even aluminum salts are suspected to be harmful to human and living things [21]. Consequently, a coagulant that is safer and produces more reusable coagulated sludge could offer a novel solution to many environmental and economic problems associated with sludge handling. However, reports on novel electrodes materials remain very scarce in the literature for the generation of reusable and environmentally friendly coagulant. Removal of metal contaminants by the chemically synthesized, different forms of, titanium dioxide was widely reported [22-25] and was similar to that of the most widely used iron and aluminum salt flocculation. Furthermore, long-term toxicological studies have not found titanium salt in water to have any adverse effects. All the above factors suggest that the titanium salt can be used as an alternative coagulant [26].

In this investigation, titanium was used instead of iron and aluminum as a novel alternative sacrificial anode, and the removal of manganese from water by titanium-based electrocoagulation was investigated. To optimize the maximum removal efficiency of manganese, different parameters like current density, pH, and temperature, inter electrode distance and co-existing ions were studied. In doing so, the equilibrium adsorption behavior is analyzed by fitting models of Langmuir, Freundlich and Redlich Peterson. The adsorption kinetics was modeled by first- and second- order rate models. Activation energy is evaluated to study the nature of adsorption.

Experimental

Chemicals

Manganese nitrate [$Mn(NO_3)_2$] of analytical grade was purchased from MERCK. Hydrochloric acid (HCl) and sodium hydroxide (NaOH) used for pH adjustment were of analytical grade from MERCK. Sodium chloride (NaCl) used for better conductivity of electrolyte of analytical grade from MERCK. Sodium phosphate, sodium silicate, sodium carbonate and sodium fluoride used as co-existing ions were of analytical grade and purchased from MERCK.

Electrolytic system and electrolysis

The experiments were carried out in a monopolar batch reactor using 1000 mL Plexiglas vessel that was fitted with a polycarbonate cell cover with slots to introduce the electrodes, pH sensor, a thermometer and the electrolytes. Titanium (Alfa Aesar, UK) of surface area (0.02 m^2) acted as the anode. The cathode was galvanized iron (commercial grade, India) sheets of the same size as the anode is placed at an inter-electrode distance of 3 cm. The temperature of the electrolyte was controlled to the desired value with a variation of ± 2 K by adjusting the rate of flow of thermo-statically controlled water through an external glass-cooling spiral. A regulated direct current (DC) was supplied from a rectifier (10 A, 0-25 V; Aplab model).

The required concentration of manganese was prepared using Milli Q water. In all the experiments 3 g L^{-1} of sodium chloride was used for better conductivity. The solution volume of 900 mL was used for each experiment as the electrolyte. The pH of the electrolyte was adjusted and measured initially and during the electrolysis by a pH meter (DKK-TOC, Japan). The pH was adjusted using either 0.1 M NaOH or 0.1 M HCl as necessary. After adjusting the initial solution pH to the desired value (3 to 9), the current density was set. The solution was stirred at 250 rpm to ensure good mixing and transport of reactants. Temperature studies were carried at varying temperature (323-343 K) to determine the type of reaction.

Analytical procedures

The concentration of manganese was determined using UV-visible Spectrophotometer with manganese kits (MERCK, Pharo 300, Germany). The SEM image of titanium dioxide was analyzed with a Scanning Electron Microscope (SEM) made by Hitachi (model s-3000h). The constituents of the titanium dioxide were analyzed by X-Ray Fluorescence (XRF) made by Horiba (model XGT-2700). The Fourier transform infrared spectrum of titanium dioxide was obtained using Nexus 670 FTIR spectrometer (Thermo Electron Corporation, USA) and X-ray diffraction (XRD) patterns of titanium dioxide was analyzed using an X'per PRO X-ray diffractometer (PANalytical, USA). TGA of titanium dioxide was carried out in the Thermal Analyzer (TA Instruments; Model SDT Q600). The concentration of carbonate, silicate, and phosphate were determined using UV-Visible spectrophotometer with respective standard ion kits supplied by MERCK (MERCK, Pharo 300).

Results and Discussion

Effect of current density on the removal efficiency

The current density is one of the prominent factors which strongly influence the performance of electrodissolution process. The current density not only determines the coagulant dosage and bubble production rate but also the size and growth of the flocks, which can influence the treatment efficiency. Therefore, the effect of current density on the removal of manganese was investigated. Applying a constant current to the titanium effectively dissolved Ti according to $Ti \rightarrow Ti^{2+} \rightarrow TiO^{2+}$. TiO^{2+} combined easily with OH^- to form $TiO_2 \cdot H_2O$ or $Ti(OH)_4$. $Ti(OH)_4$ is unstable substance, which changes gradually into $TiO_2 \cdot H_2O$ by dehydration. The reaction equations were,

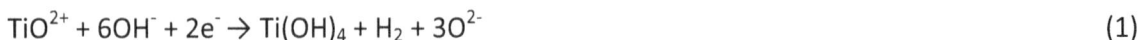

$$TiO^{2+} + 6OH^- + 2e^- \rightarrow Ti(OH)_4 + H_2 + 3O^{2-} \tag{1}$$

and

$$TiO^{2+} + 2OH^- \rightarrow TiO_2 . H_2O \tag{2}$$

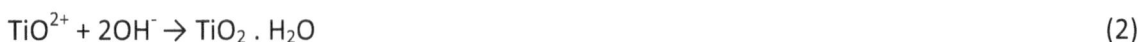

and at the cathode the following reaction is taking place,

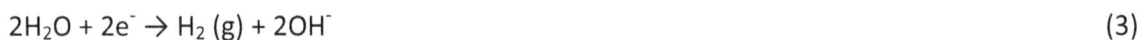

$$2H_2O + 2e^- \rightarrow H_2 (g) + 2OH^- \tag{3}$$

The amount of manganese removal depends upon the quantity of adsorbent (hydrous titanium dioxide) generated, which is related to the time and current density [27]. The amount of adsorbent was determined from Faraday's law. With the increase in current density the amount of hydrous titanium dioxide generation also increases. To investigate the effect of current density on the manganese removal, a series of experiments were carried out by solutions containing a constant pollutants loading of 2 mg L^{-1}, at a pH 7.0, with current density being varied from 0.02 to 0.1 A dm^{-2}. The removal efficiencies are 60.25, 82.54, 90.10, 97.55 and 97.90 % for 0.02, 0.04, 0.06, 0.08 and 0.1 A dm^{-2} respectively. From the results, it was found that very small raise in removal efficiency was observed for current densities 0.08 and 0.1 A dm^{-2}. Hence, the further experiments were carried out at 0.08 A dm^{-2}.

Effect of pH on the removal efficiency

It is believed that the initial pH is an important operating factor influencing the performance of electrodissolution process. To explain this effect, a series of experiments were carried out using 2 mg L^{-1} of manganese containing solutions, by adjusting the initial pH in the interval from 3 to 9. The removal efficiencies for the pH 3, 4, 5, 6, 7, 8 and 9 are 60, 79, 82.5, 95.50, 97.55, 97.56 and 97.65 % respectively. It is well know that the titanium dioxide adsorption is pH dependent. At acidic pH it is positively charged while at alkaline pH it negatively charged. Mostly point of zero charge for titanium dioxide is approximately pH 6 to 8 [28,29]. The results agreed well with earlier results from the literature. Further experiments were carried out at pH 7.

Effect of electrolyte concentration

In order to evaluate the effect of initial concentration of manganese, experiments were conducted with varying initial concentration from 0.25-2.0 mg L^{-1}. Figure 1 shows that the uptake of manganese (mg g^{-1}) increased with increase in manganese concentration and remained nearly constant after equilibrium time. The equilibrium time was found to be 180 min for all concentration studied. The amount of manganese adsorbed (q_e) increased from 0.248 to 1.982 mg g^{-1} as the concentration increased from 0.25-2.0 mg L^{-1}. From the Figure 1 it is found that the plots are single, smooth and continuous curves leading to saturation, suggesting the possible monolayer coverage to manganese on the surface of the adsorbent.

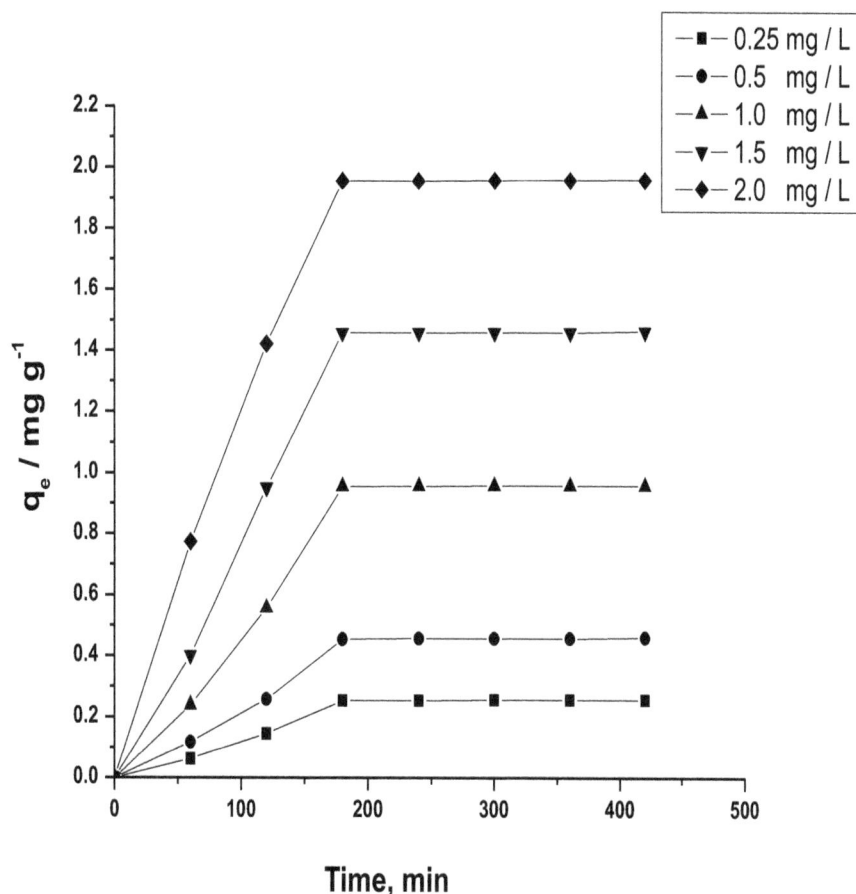

Figure 1. *Effect of time and initial concentration of manganese for the adsorption on hydrous titanium dioxide, pH 7.0, T = 303K.*

Effect of competing ions

Carbonate

Effect of carbonate on manganese removal was evaluated by increasing the carbonate concentration from 0 to 250 mg L^{-1} in the electrolyte. The removal efficiencies are 97.55, 95.3, 72.8, 50.7, 38, and 19 % for the carbonate concentration of 0, 2, 5, 65, 150 and 250 mg L^{-1}, respectively. From the results it is found that the removal efficiency of the manganese is not affected by the presence of carbonate below 2 mg L^{-1}. Significant reduction in removal efficiency was observed above 5 mg L^{-1} of carbonate concentration is due to the passivation of anode resulting, the hindering of the dissolution process of anode [30].

Phosphate

The concentration of phosphate ion was increased from 0 to 50 mg L^{-1}, the contaminant range of phosphate in the ground water. The removal efficiency for manganese was 97.55, 91.3, 60.7, 35.5 and 29.2 % for 0, 2, 5, 25 and 50 mg L^{-1} of phosphate ion, respectively. There was no change in removal efficiency of manganese below 2 mg L^{-1} of phosphate in the water. At higher concentrations (at and above 5 mg/L) of phosphate, the removal efficiency decreased drastically. This was due to the preferential adsorption of phosphate over manganese as the concentration of phosphate increased.

Arsenic

The concentration of arsenic was gradually increased from 0 to 5 mg L^{-1}. From the results it was found that the efficiency decrease for manganese was 97.55, 90.7, 78.5, 68.6 and 44.6 % by increasing the concentration of arsenate from 0, 0.2, 0.5, 2.5 and 5.0 mg L^{-1}, respectively. This was due to the preferential adsorption of arsenic over manganese as the concentration of arsenate increases. So, when arsenic ions are present in the water to be treated arsenic ions compete greatly with manganese ions for the binding sites.

Silicate

From the results it is found that no significant change in manganese removal was observed, when the silicate concentration was increased from 0 to 2 mg L^{-1}. The respective efficiencies for 0, 2, 5, 10 and 15 mg L^{-1} of silicate are 97.55, 80.2, 72.4, 51.6 and 43.8 %. In addition to preferential adsorption, silicate can interact with titanium dioxide to form soluble and highly dispersed colloids that are not removed by normal filtration [30].

Adsorption kinetic modeling

The kinetic studies predict the progress of adsorption; however, the determination of the adsorption mechanism is also important for design purposes. In this research investigation, first- and second order kinetic models were tested at different concentration (0.25 to 2.0 mg L^{-1}) at a current density of 0.08 A dm^{-2}.

First order kinetic model

The first order kinetic model is generally expressed as follows [31],

$$dq_t/dt = k_1 (q_e\text{-}q_t) \tag{4}$$

where q_e / mg g^{-1} and q_t / mg g^{-1} are the adsorption capacities at equilibrium and at time t / min respectively, and k_1/ min^{-1} is a rate constant of first order adsorption. The integrated form of the above equation with the boundary conditions $t = 0$ to $t = t$ and $q_t = 0$ to $q_t = q_t$ is rearranged to obtain the following time dependence function,

$$\log(q_e\text{-}q_t) = \log q_e - k_1 t / 2.303 \tag{5}$$

The experimental data were analyzed initially with first order model. The plot of log (q_e-q_t) vs. t should give the linear relationship from which k_1 and q_e can be determined by the slope and intercept, respectively Eq. (5). The computed results are presented in Table 1. The results show that the theoretical q_e (cal) value doesn't agree to the experimental q_e (exp) values at all concentrations studied with poor correlation coefficient. This result indicated that the adsorption system do not follow a first-order reaction. So, further the experimental data were fitted with second order model.

Second order kinetic model

The second order kinetic model is expressed as [32],

$$dq_t/dt = k_2 (q_e\text{-}q_t)^2 \tag{6}$$

The integrated form of Eq. (6) with the boundary condition $t = 0$ to $t = t$ and $q_t = 0$ to $q_t = q_t$ is,

$$1/(q_e\text{-}q_t) = 1/q_e + k_2 t \tag{7}$$

Eq. (7) can be rearranged and linearized as,

$$t/q_t = 1/k_2 q_e^2 + t/q_e \qquad\qquad (8)$$

where, q_e / mg g^{-1} and q_t / mg g^{-1} are the amount of manganese adsorbed on hydrous titanium dioxide at equilibrium and at time t / min, respectively, and k_2 is the rate constant for the second order kinetic model.

Table 1 *Comparison between the experimental and calculated q_e values at different concentrations in first order and second order adsorption kinetics at a current density of 0.08 A dm^{-2}.*

C / mg L^{-1}	q_e / mg g^{-1} (exp)	Pseudo first order adsorption			Pseudo Second order adsorption		
		q_e / mg g^{-1} (cal)	k_1 / min^{-1}	R^2	q_e / mg g^{-1} (cal)	k_2 / (g mg^{-1}) min^{-1}	R^2
0.25	0.248	35.78	0.0244	0.0069	0.248	1.6715	0.992
0.50	0.463	22.84	0.0212	0.0204	0.461	0.1013	0.991
1.0	0.951	1.543	0.0168	0.6154	0.950	0.1071	0.995
1.5	1.470	1.337	0.0124	0.3462	1.470	0.0874	0.986
2.0	1.982	5.780	0.0178	0.3425	1.980	0.0343	0.987

The kinetic data were fitted to the second order model Eq. (8). The equilibrium adsorption capacity, q_e (cal) and k_2 were determined from the slope and intercept of plot of t/q_t versus t and are compiled in Table 1. Figure 2 shows the plot of t/q_t versus t for manganese adsorption and the plots were found to be linear. The theoretical q_e (cal) value also agreed very well with the experimental q_e value, indicating the pseudo second-order kinetics. In addition, the correlation coefficient for the second-order kinetic model was 0.99, which suggest the applicability of this kinetic equation and the second-order nature of the sorption process of manganese on hydrous titanium dioxide.

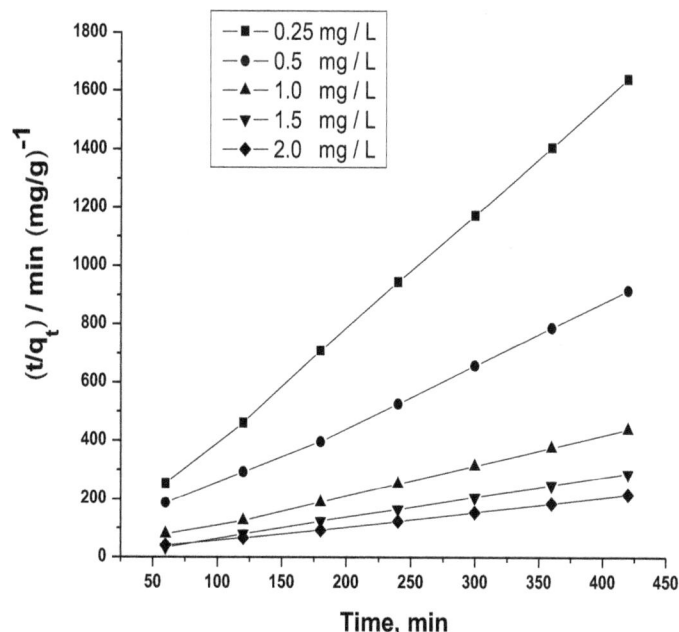

Figure 2. *Second-order kinetic model plots for adsorption of manganese at different concentrations, pH of the electrolyte: 7.0, temperature: 303 K, current density: 0.08 A dm^{-2}*

The computed results obtained from first order and second order models were depicted in Table 1. From the tables, it was found that the correlation coefficient values are in the order of second order > first order. This indicates that the adsorption follows the second order model. Further, the calculated q_e values well agree with the experimental q_e values for second order kinetics model. These results indicate that the second-order kinetic model can be applied suitably to predict the manganese adsorption process onto hydrous titanium dioxide.

Isotherm modeling

In order to explain the mechanism of the adsorption process, it is important to establish the most appropriate correlation for the equilibrium curves. In this study, three adsorption isotherms viz., Freundlich, Langmuir and Redlich isotherm models were applied to establish the relationship between the amounts of manganese adsorbed onto the hydrous titanium hydroxide and its equilibrium concentration in the electrolyte containing contaminant ions.

Freundlich Isotherm

The Freundlich adsorption isotherm typically fits the experimental data over a wide range of concentrations. This empirical model includes considerations of surface heterogeneity and exponential distribution of the active sites and their energies. The isotherm is adopted to describe reversible adsorption and is not restricted to monolayer formation. The linearised in logarithmic form and the Freundlich constants can be expressed as [33],

$$\log q_e = \log k_f + n \log C_e \tag{9}$$

where, k_f is the Freundlich constant related to adsorption capacity, n is the energy or intensity of adsorption, C_e is the equilibrium concentration of manganese (mg L^{-1}).

In testing the isotherm, the manganese concentration used was 0.25 to 2.0 mg L^{-1}, current density of 0.08 A dm^{-2} and at an initial pH 7. The adsorption data is plotted as log q_e versus log C_e by equation (9) should result in a straight line with slope n and intercept k_f. The intercept and the slope are indicators of adsorption capacity and adsorption intensity, respectively. The value of n falling in the range of 1-10 indicates favorable sorption. Freundlich constant (k_f) and n values were listed in Table 2. From the analysis of the results it is found that the Freundlich plots fit only satisfactorily with the experimental data obtained in the present study which is shown in the Figure. 3 (a).

Table 2 *Constant parameters and correlation coefficient for different adsorption isotherm models for manganese adsorption at 0.25- 2.0 mg L^{-1} at a current density of 0.08 A dm^{-2}*

Isotherm	Parameters	Concentration of Mn, mg L^{-1}				
		0.25	0.5	1.5	1.5	2.0
Langmuir	q_m / mg g^{-1}	0.2383	0.4597	0.9564	1.4616	1.9491
	b / L mg^{-1}	0.1113	0.1102	0.1099	0.1014	0.0948
	R^2	0.9943	0.9987	0.9954	0.9962	0.9991
	R_L	0.9729	0.9479	0.8998	0.8569	0.8179
Freundlich	k_f / mg g^{-1}	0.5803	0.5512	0.5174	0.4897	0.4613
	n / L mg-1)	2.1786	2.0257	1.9457	1.8798	1.7259
	R^2	0.9812	0.9789	0.9881	0.9836	0.9820
Redlich Peterson	K_F / L g^{-1}	0.9978	0.9981	0.9968	0.9990	0.9891
	β	0.9764	0.9854	0.9817	0.9897	0.9789
	a_R / L mmol$^{-(1-1/\beta)}$	27.412	28.417	29.648	30.568	32.516

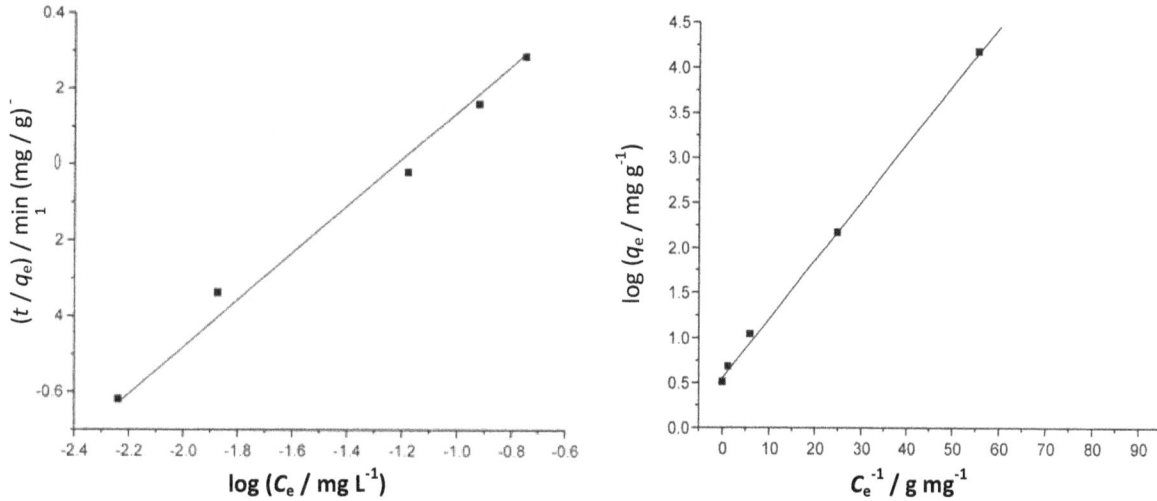

Figure 3. (a) *Frendlich plot (log q_e vs log C_e) for adsorption of manganese, pH of the electrolyte: 7.0, current density: 0.08 A dm^{-2}, concentration: 2.0 mg L^{-1}, **(b)** Langmuir plot (1/qe vs. 1/Ce) for adsorption of manganese, pH of the electrolyte: 7.0, current density: 0.08 A dm^{-2}, concentration: 2.0 mg L^{-1}*

Langmuir Isotherm

This model assumes a monolayer deposition on a surface with a finite number of identical sites. It is well known that the Langmuir equation is valid for a homogeneous surface. The linearized form of Langmuir adsorption isotherm model is [34],

$$C_e/q_e = 1/q_m b + C_e/q_m \tag{10}$$

where, q_e is amount adsorbed at equilibrium, C_e is the equilibrium concentration, q_m is the Langmuir constant representing maximum monolayer adsorption capacity and b is the Langmuir constant related to energy of adsorption. The essential characteristics of the Langmuir isotherm can be expressed as the dimensionless constant R_L.

$$R_L = 1 / (1 + bC_o) \tag{11}$$

where R_L is the equilibrium constant it indicates the type of adsorption, b, is the Langmuir constant. C_o is various concentration of manganese solution. The R_L values between 0 and 1 indicate the favorable adsorption.

Langmuir isotherm was tested from Eq. (10). The plots of $1/q_e$ as a function of $1/C_e$ for the adsorption of manganese on hydrous titanium dioxide are shown in Figure 3 (b). The plots were found linear with good correlation coefficients (>0.99) indicating the applicability of Langmuir model in the present study. The values of monolayer capacity (q_m) and Langmuir constant (b) is presented in Table 2. The values of q_m calculated by the Langmuir isotherm were all close to experimental values at given experimental conditions. These facts suggest that manganese is adsorbed in the form of monolayer coverage on the surface of the adsorbent. The dimensionless constant R_L was calculated from Eq.(11). The R_L values were found to be between 0 and 1 for all the concentration of manganese studied. The correlation co-efficient values of Langmuir and Freundlich isotherm models are presented in Table 2.

Redlich Peterson isotherm

It is a three parameter hybrid isotherm. It is having features of both the Langmuir and Frendlich isotherms. This model has linear dependence in the numerator component and exponential component in denominator of non-linear form [35,36].

$$q_e = \frac{K_F C_e}{1 + a_R C_e^{\beta}}$$
(12)

where q_e / mmol g^{-1} is the solid-phase sorbate concentration at equilibrium, C_e / mmol L^{-1} is the concentration of adsorbate in equilibrium with liquid phase, K_F / L g^{-1} and a_R / L mmol$^{-(1-1/\beta)}$ are the Redlich-Peterson isotherm constants, and β is the exponent, which lies between 1 and 0. If the β tends to 0 then the adsorption follows the Frendlich isotherm and if the β value tends to one it fits with the Langmuir isotherm. In order to verify our investigation regarding the monolayer or multilayer adsorption, the linear form of Redlich-Peterson is used. It is little bit complicated compared to the other isotherms. The strategy to find these parameters is based in the maximization of correlation coefficients (R^2) from the linear fit to the data. In this way the K_F values are modified until obtain the best fit of the data. The linear form of the equation for this model is

ln ($K_F(C_e/q_e-1)$) = ln a_R + β ln C_e
(13)

Plotting of ln ($K_F(C_e/q_e-1)$ vs. ln C_e by the Eq. (13) gives the Redlich Peterson equation. This isotherm is a three parameters isotherm in which K_F values are indirectly obtained by plotting the graph with maximum correlation coefficient by justifying the values of K_F. By that K_F, β and a_R are in the Table 2 for all concentrations. Here the β values are above the 0.95. So the adsorption favors Langmuir isotherm rather than Frendlich isotherm.

Adsorption thermodynamics

To understand the effect of temperature on adsorption process, thermodynamic parameters should be determined at various temperatures. The energy of activation for adsorption of manganese can be determined by the second order rate constant is expressed in Arrhenius form [37],

ln k_2 = ln k_o - E/RT
(14)

where k_o is the constant of the equation (g mg^{-1}) min^{-1}), E is the energy of activation (J mol^{-1}), R is the gas constant (8.314 J mol^{-1} K^{-1}) and T is the temperature (K). Figure 4(a) shows that the rate constants vary with temperature according to Eq.(14) giving an activation energy of 21.01 kJ mol^{-1} for manganese from the slope of the fitted equation. The free energy change is obtained using the following relationship

ΔG = -RT ln K_c
(15)

where ΔG is the free energy (kJ mol^{-1}), K_c is the equilibrium constant, R is the universal gas constant and T is the temperature in K. The values of K_c and ΔG are presented in Table 3. The negative value of ΔG indicates the spontaneous nature of adsorption. Other thermodynamic parameters such as entropy change (ΔS) and enthalpy change (ΔH) were determined using the van't Hoff equation:

$$\ln K_c = \frac{\Delta S}{R} - \frac{\Delta H}{RT}$$
(16)

The enthalpy change (ΔH = -60.57 J mol^{-1}) and entropy change (ΔS = -0.047 J mol^{-1} K^{-1}) were obtained from the slopes and intercepts of the van't Hoff linear plots of ln K_c versus $1/T$ (Figure. 4(b)) Eq.(16). Negative value of enthalpy change (ΔH) indicates that the adsorption process is exothermic in nature, and the negative value of change in internal energy (ΔG) show the spontaneous adsorption of manganese on the adsorbent. Negative values of entropy change show the increased randomness of the solution interface which gains heat from the surroundings during adsorption of manganese on the adsorbent [38] (Table 3). Negative enthalpy and negative entropy shows that the adsorption is more favorable at low temperature. This is due to the decrement of pore size as temperature increases.

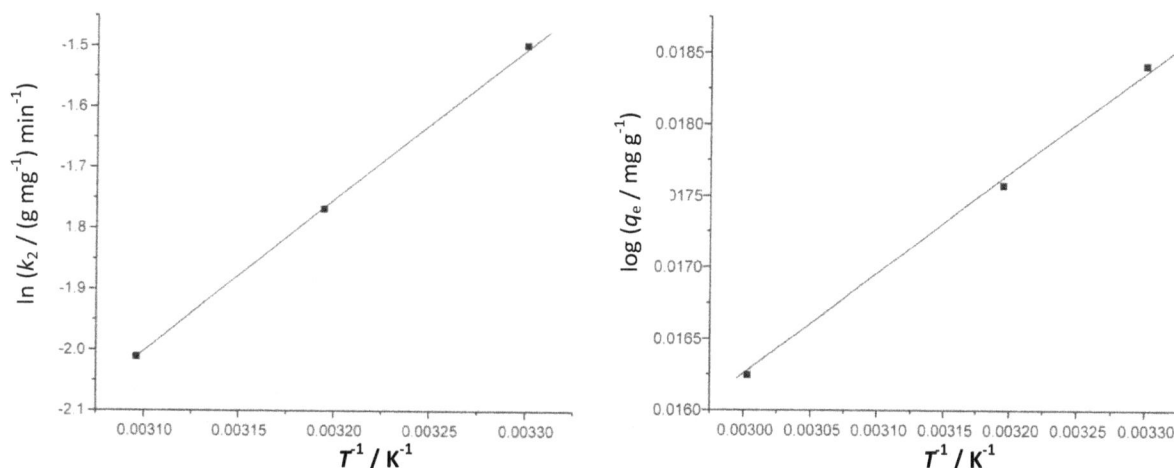

Figure 4. (a) Plot of log k_2 vs. T^{-1}; **(b)** Plot of ln K_c vs. T^{-1}: pH 7.0; j = 0.08 A dm^{-2}, C = 2 mg L^{-1}

The enthalpy change (ΔH = -60.57 J mol^{-1}) and entropy change (ΔS = -0.047 J mol^{-1} K^{-1}) were obtained from the slopes and intercepts of the van't Hoff linear plots of ln K_c versus 1/T (Figure. 4(b)) Eq.(16). Negative value of enthalpy change (ΔH) indicates that the adsorption process is exothermic in nature, and the negative value of change in internal energy (ΔG) show the spontaneous adsorption of manganese on the adsorbent. Negative values of entropy change show the increased randomness of the solution interface which gains heat from the surroundings during adsorption of manganese on the adsorbent [38] (Table 3). Negative enthalpy and negative entropy shows that the adsorption is more favorable at low temperature. This is due to the decrement of pore size as temperature increases.

Table 3 Thermodynamics parameters for adsorption of manganese.

Temperature, K	K_c	ΔG^o / kJ mol^{-1}	ΔH^o / Jmol^{-1}	ΔS^o / J mol^{-1} K^{-1}
323	1.0433	-0.0464		
333	1.0406	-0.0450	-60.57	-0.0470
343	1.0382	-0.0436		

Table 4. Comparison between the experimental and calculated q_e values at different temperatures in first and second order adsorption kinetics of manganese: C = 2.0 mg L^{-1}, pH 7.0, j = 0.08 A dm^{-2}

T / K	q_e / mg g^{-1} (exp)	First order adsorption			Second order adsorption		
		q_e / mg g^{-1} (cal)	k_1 / min^{-1}	R^2	q_e / mg g^{-1} (cal)	k_2 / (g mg^{-1}) min^{-1}	R^2
323	2.048	1.112	-0.0061	0.5430	2.01	0.0320	0.990
333	2.160	0.987	-0.0068	0.4569	2.13	0.0095	0.986
343	2.165	0.874	-0.0071	0.4037	2.14	0.0097	0.991

Using Lagergren rate equation, pseudo second order rate constants and correlation co-efficient were calculated for different temperatures (323-343 K). The calculated q_e values obtained from the second order kinetics agrees with the experimental q_e values better than the first order kinetics model, indicating adsorption following second order kinetics. Table 4 depicts the computed results obtained from pseudo first and pseudo second order kinetic models.

Characterization of hydrous titanium dioxide

XRF studies

The contents of titanium dioxide were analyzed by XRF. The titanium dioxide sample was dried in a drying chamber at 100 °C were ground in the agate mortar. As shown in Table 5, 95.2 % of the sample, by weight, was titanium dioxide. Sodium chloride was came from the electrolyte, and calcium and barium elements came from the impurity in the titanium electrode.

XRD studies

The crystal structure of hydrous titanium dioxide nanoparticles was analyzed by X-ray powder diffractometer operating with CuKα radiation source filtered with a graphite monochromator. Figure. 5(a) shows the X-ray diffraction pattern of hydrous titanium dioxide nanoparticles. From the figure it is found that, most diffraction peaks belong to the anatase phase (JCPDS Card Number 73-1764), and minor peaks from the brookite phase (JCPDS Card Number 76-1936) could also be observed. The crystallite size D was determined from the broadening of corresponding strongest X-ray diffraction peaks by using Scherrer's formula [39]:

$$D = \frac{0.9\lambda}{\beta \cos\theta} \tag{17}$$

where D is the crystalline size, λ is the average wavelength of the X-ray radiation (λ = 1.5418 Å), β is the line-width at half-maximum peak position, and θ is the diffracting angle (2θ = 25.4°). The average crystallite size of the hydrous titanium dioxide is 4.3-8.4 nm.

FT-IR spectrum

The hydrous titanium dioxide was analyzed using FTIR and results are presented in Figure 5(b). The strong peak at 3381 cm^{-1} is attributed to the stretching vibrations of surface and interlayer water molecules and hydroxyl groups. This is related to the formation of hydrogen bonds of inter-layer water with guest anions as well as with hydroxide groups of layers. At 1630.18 cm^{-1}, there is a strong adsorption peak for hydroxyl bending vibration belonging to physically adsorbed H_2O. One small adsorption peak could also be identified at 1371.37 cm^{-1}, which represents the coordinated hydroxyl groups. These observations demonstrate that these hydrous titanium dioxide nanoparticles have high adsorption capacities to H_2O and hydroxyl groups exist on their surfaces [40].

SEM and EDAX analysis

Figure 5(c) shows the SEM images of hydrous titanium dioxide. The SEM images show different size, shape and dimension and these nanoparticles are aggregated into micro-sized particles. The volume median diameter value of these nanoparticles in distilled water was determined at approximately 9.8 nm by the dynamic light scattering technique, which is in accordance with the SEM observation. This type of aggregation of nanoparticles is beneficial to their removal from aqueous environment after the treatment process.

Energy-dispersive analysis of X-rays was used to analyze the elemental constituents of titanium dioxide generated during the electro dissolution process and the results are presented in Figure 5(d). The figure indicates that the titanium dioxide was composed mainly of Ti an O, which affirms that the titanium dioxide was generated by anodic dissolution.

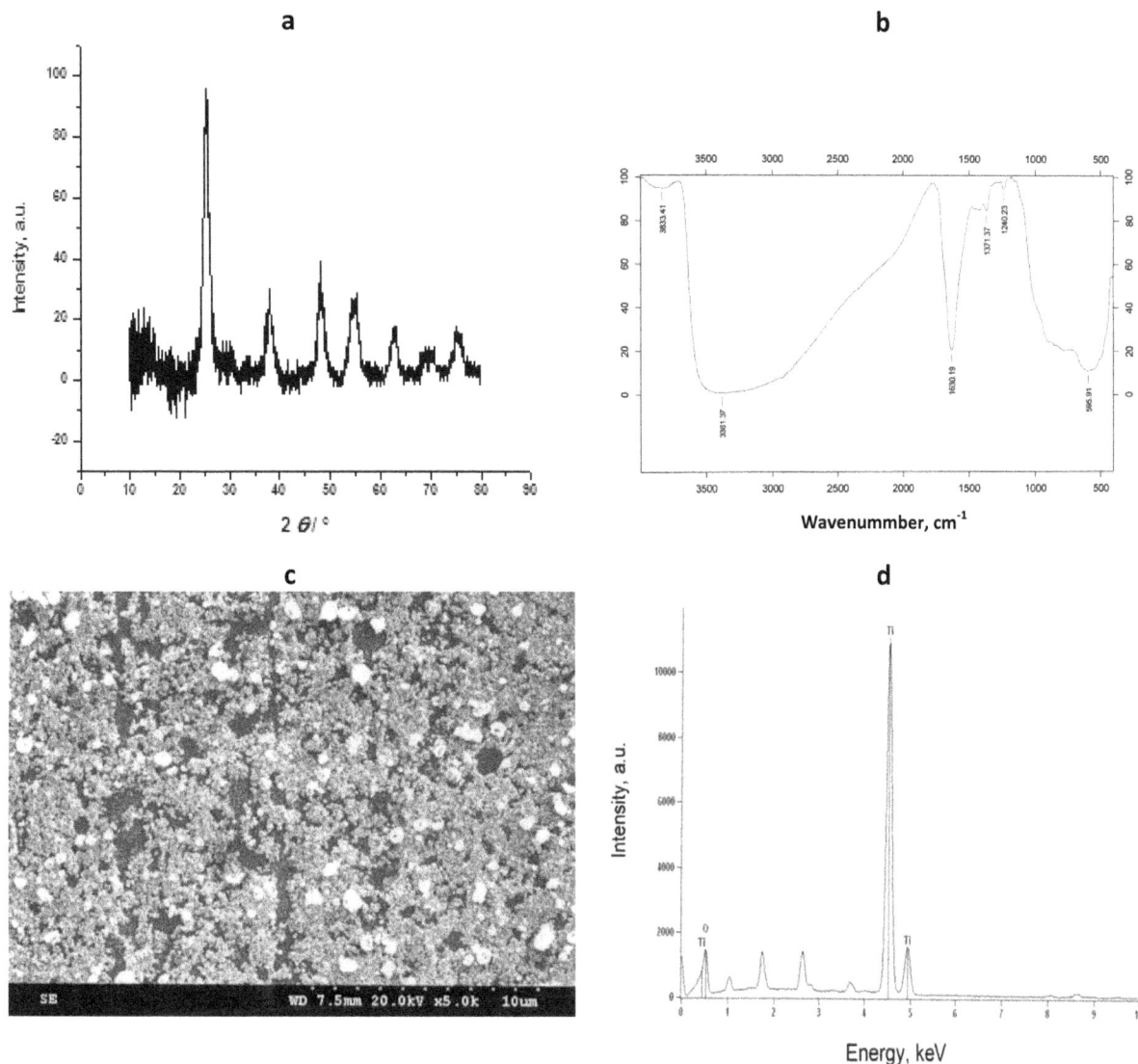

Figure 5. (a) X-ray diffraction pattern of hydrous TiO_2, **(b)** FTIR pattern of hydrous TiO_2, **(c)** SEM image of the hydrous TiO_2, **(d)** EDAX image of the hydrous TiO_2

TGA analysis

TGA analysis (figure not shown) of hydrous titanium dioxide was carried out. From the results we found that the weight loss of 13.0 % where observed when the samples were heated from the 32 - 800°C. The entire range will be divided into three stages *viz.*, first, second and third stage. In the first stage (32 – 122 °C) weight loss (6.9%) could be attributed to the elimination of physically absorbed water. In the second stage (122 to 438 °C) weight loss (6.0 %) could be contributed to the loss of surface hydroxyl groups. In the third stage (438 to 800 °C) no exothermic peak was observed and the weight loss is around 0.1 %.

Conclusions

The maximum removal efficiency of 97.55 % was achieved with titanium as sacrificial anode at a current density of 0.08 A dm^{-2}, pH 7.0. The results indicate that the hydrous titanium dioxide, by electro-dissolution of sacrificial anodes, efficiently adsorbs the manganese from water. Hence this process can be used as an effective process for the removal of manganese contaminated water resources. The results indicate that, the second-order kinetic model accurately described the adsorption kinetics. The adsorption mechanism was found to be chemisorption and the rate-limiting step was mainly surface adsorption. The Langmuir isotherm showed a better fit than the Freundlich and Redlich isotherms, thus, indicating the applicability of monolayer coverage of manganese on hydrous titanium dioxide.

The thermodynamic parameters like ΔG, ΔH and ΔS were determined. Their values indicated that the adsorption process was favorable, spontaneous, and exothermic in nature. As the temperature increased ΔG became less negative, indicating a stronger driving force, resulting in a greater adsorption capacity at higher temperatures. The negative value of ΔH confirmed that the process was exothermic. Negative values of entropy change show the increased randomness of the solution interface which gains heat from the surroundings during adsorption of manganese on the adsorbent. EDAX analysis confirmed that manganese was adsorbed on to the hydrous titanium dioxide.

Acknowledgments: *The authors wish to express their gratitude to Dr. Vijayamohanan K. Pillai, Director, CSIR-Central Electrochemical Research Institute, Karaikudi to publish this article.*

References

[1]   Y. H. Chang, K. H. Hsieh, F. C. Chang, *Journal of Applied Polymer Science* **112** (2009) 2445–2454.

[2]   K. Kannan, Fundamentals of Environmental Pollution, S Chand Co. Limited, New Delhi, 1995

[3]   Y. C. Sharma Uma, S. N. Singh Paras, F. Gode, *Chemical Engineering Journal* **132** (2007) 319–323.

[4]   A. Takeda, *Brain Research Reviews* **41** (2003) 79–87.

[5]   J. Donaldson, *Neuro Toxicology* **8** (1987) 451–462.

[6]   S. M. Bamforth, D. A.C. Manning, I. Singleton, P. L. Younger, K. L. Johnson, *Applied Geochemistry* **21** (2006) 1274–1287

[7]   R. W. Leggett, *Science of the Total Environment* **409** (2011) 4179–4186

[8]   M. K. Doula, *Water Research* **40** (2006) 3167-3176.

[9]   World Health Organization, *Manganese in drinking water*, Background document for development of WHO guidelines for drinking water quality report: 2011. ([www.who.int/ /water_sanitation_health/dwq/chemicals/manganese.pdf](www.who.int/)), accessed November 2014

[10] S. R. Taffarel, J. Rubio, *Minerals Engineering* **22** (2009) 336–343

[11] M. K. Doula, *Water Research* **40** (2006) 3167-3176

[12] D. Barloková, J. Ilavský, *Polish Journal of Environmental Studies* **19** (2010) 1117-1122

[13] S.-C. Han, K.-H. Choo, S.-J. Choi, M. M. Benjamin, *Journal of Membrane Science* **290** (2007) 55–61

[14] A. G. Tekerlekopoulou, D. V. Vayenas, *Desalination* **210** (2007) 225–235

[15] E. Okoniewska, J. Lach, M. Kacprzak, E. Neczaj, *Desalination* **206** (2007) 251–258

[16] A. Omri , M. Benzina, *Alexandria Engineering Journal* **51** (2012) 343–350

[17] D. Simonsson, *Chemical Society Reviews* 26 (1997) 181–189.

[18] G. Chen, *Separation &.Purification Technology* **38** (2004) 11 – 41

[19] S. Vasudevan, M. A. Oturan, *Environmental Chemistry Letters* **12** (2014) 97 – 108

[20] M. Ben Sasson, A. Adin, *Water Research* **44** (2010) 3973 – 3981.

[21] W. P. Cheng, F. H. Chi, *Water Research* **36** (2002) 4583–4591.

[22] M. Patel, L. Lippincott, X. Meng, *Water Research* **39** (2005) 2327-2337

[23] M. Pirilä, M. Martikainen, K. Ainassaari, T. Kuokkanen, R. L. Keiski, *Journal of Colloid and Interface Science* **353** (2011) 257–262

[24] Z. Xu, Q. Li, S. Gao, J. K. Shang, *Water Research* **44** (2010) 5713-5721

[25] M. C. Lu, G..D. Roam, J..N. Chen, C. P. Huang, *Water Research* **30** (1996) 1670-1678

[26] H. K. Shon, S. Vgneswaran, I. S. Kim, J. Cho, G. J. Kim, J. B. Kim, J. H. Kim, *Environmental Science and Technology* **42** (2007) 1372-1377

[27] S. Vasudevan, J. Lakshmi, G. Sozhan, *Desalination* **310** (2013) 122-129

[28] P. Westerhoff, *Arsenic Removal with Agglomerated Nanoparticle Media*, AWWA Research Foundation, Arizona state University, 2006.

[29] H. Jezequel, K. H. Chu, *Journal of Environmental Science Health A.* **41**(2006) 1519-1528.

[30] S. Vasudevan, J. Lakshmi, J. Jayaraj, G. Sozhan, *Journal of Hazardous Materials* **164** (2009) 1480-1486.

[31] [Y. P. Teoh, M. Ali Khan, T. S. Y. Choong, *Chemical. Engineering Journal* **217** (2013) 248–255

[32] Y. S. Ho, G. McKay, *Process Biochemistry* **34** (1999) 451 – 456.

[33] H. M. F. Freundlich, *Journal of Physical Chemistry* **57** (1906) 385 – 470.

[34] I. Langmuir, *Journal of American Chemical Society.* **40** (1918) 1316-1403.

[35] Y. S. Ho, J. F. Porter, G. Mckay, *Water Air. Soil Pollution* **141** (2002) 1-33.

[36] K. Y. Foo, B. H. Hameed, *Chemical Engineering Journal* **156** (2010) 2-10.

[37] S. Pan, H. Shen, Q. Xu, J. Luo, M. Hu, *Journal of Colloid. Interface Science* **365** (2012) 204–212.

[38] L. D. Rio, J. Aberg, R. Renner, O. Dahiaten , V. Vedral, *Nature* **474** (2011) 61-63

[39] C. S. Barrett, T. B. Massalski, *Structure of Metals, Third edition*, Mc Graw-Hill, NewYork, 1966

[40] S. Debnath, U. C. Ghosh, *Desalination* **273** (2011) 330-342.

15

Electrochemical investigation of the corrosion behavior of heat treated Al-6Si-0.5Mg-xCu (x=0, 0.5 and 1) alloys

Abul Hossain✉, Mohammed Abdul Gafur*, Fahmida Gulshan and
Abu Syed Wais Kurny

Department of Materials and Metallurgical Engineering, Bangladesh University of Engineering and Technology, Dhaka, Bangladesh
Pilot Plant and Process Development Centre (PP & PDC), BCSIR Laboratories, Dhaka, Bangladesh

✉Corresponding Author

Abstract

The corrosion behavior of heat treated Al-6Si-0.5Mg-xCu (x=0, 0.5 and 1 wt %) alloys in 0.1 M NaCl solution was investigated using potentiodynamic polarization and electrochemical impedance spectroscopy (EIS) techniques. The potentiodynamic polarization curves reveal that 0.5 wt % Cu and 1 wt % Cu content alloys are less prone to corrosion than the Cu free alloy. The EIS test results also showed that corrosion resistance or charge transfer resistance (R_{ct}) increases with increasing Cu content into Al-6Si-0.5Mg alloy. Maximum charge transfer resistance (R_{ct}) was obtained with the addition of 1 wt % Cu and minimum R_{ct} value was for Cu free Al-6Si-0.5Mg alloy. Due to addition of Cu and thermal modification, the magnitude of open circuit potential (OCP), corrosion potential (E_{corr}) and pitting corrosion potential (E_{pit}) of Al-6Si-0.5Mg alloy in NaCl solution were shifted to the more noble direction.

Keywords

Al alloy; Nyquist plot; Corrosion rate; Tafel plot; EIS

Introduction

Aluminium and its alloys are considered to be highly corrosion resistant under the majority of service conditions [1]. The various grades of pure aluminum are the most resistant, followed closely by the Al-Mg and Al-Mn alloys. Next in order are Al-Mg-Si and Al-Si alloys. The alloys containing copper are the least resistant to corrosion [2]; but this can be improved by coating each side of the copper containing alloy with a thin layer of high purity aluminium, thus gaining a three ply metal (Alclad). This cladding acts as a mechanical shield and offers sacrificial protection [3]. When aluminum surfaces are exposed to atmosphere, a thin invisible oxide (Al$_2$O$_3$) skin forms, which protects the metal from further corrosion in many environments [1]. This film protects the metal from further oxidation unless this coating is destroyed, and the material remains fully protected against corrosion [3]. The composition of an alloy and its thermal treatment are important do determine the susceptibility of the alloy to corrosion [4,5].

Over the years a number of studies have been carried out to assess the effect of Cu content and the distribution of second phase intermetallic particles on the corrosion behavior of Al alloys. The distribution of Cu in the microstructure affects the susceptibility to localized corrosion. Intergranular corrosion (IGC) is generally believed to be associated with Cu containing grain boundary precipitates and the precipitates free zones (PFZ) along grain boundaries [6-8]. In heat treatable Al-Si-Mg(-Cu) series alloys the susceptibility to localized corrosion (pitting and / or intergranular (IGC)) and the extent of attack are mainly controlled by the type, amount and distribution of the precipitates which form in the alloy during any thermal or thermomechanical treatment performed during manufacturing processes [6-10].

Depending on the composition of the alloy and parameters of the heat treatment process, these precipitates form in the bulk of the grain, or in the bulk as well as grain boundaries. As indicated by several authors, the precipitates formed by heat treatment in Al-Si-Mg alloys containing Cu are the θ (Al$_2$Cu) Q-phase (Al$_4$Mg$_8$Si$_7$Cu$_2$), β-phase (Mg$_2$Si) and free Si if Si content in the alloy exceeds the Mg$_2$Si stoichiometry [2-12].

The present study is an attempt to investigate the corrosion behavior of Al-6Si-0.5Mg alloys containing 0.5 and 1 wt % Cu in 0.1M NaCl solution and to examine corroded surfaces by optical and scanning electron microscopy.

Experimental

Materials preparation: The Al-6Si-0.5Mg-xCu(x= 0, 0.5 and 1) alloys were prepared by melting Al-7Si-0.3Mg (A356) alloys and adding Al and Cu into the melt. The melting operation was carried out in a gas fired clay graphite crucible furnace and the alloys were cast in a permanent steel mould. After solidification the alloys were homogenised (500 °C for 24 hr), solution treated (540 °C for 2 hr) and finally artificially aged (225 °C for 1 hr). After heat treatment rectangular samples (30x10x5 mm) were prepared for metallographic observation and subsequent electrochemical test. Deionized water and analytical reagent grade sodium chloride (NaCl) were used for the preparation of 0.1 M solution. All measurements were carried out at room temperature.

Potentiodynamic polarization measurements: A computer-controlled Gamry Framework TM Series G 300™ and Series G 750™ Potentiostat/Galvanostat/ZRA were used for the electrochemical measurements. The potentiodynamic polarization studies were configured in cells, using three-electrode assembly: a saturated calomel reference electrode, a platinum counter electrode and the sample in the form of coupons of exposed area of 0.50 cm^2 or 10 x 5 mm as working electrode. Only one 10x5 mm surface was exposed to the test solution, the other surfaces being covered with Teflon tape. The system was allowed to establish a steady-state open circuit potential (OCP). The potential range selected was -1 to +1V and measurements were made at a scan rate of 0.50 mV/s. The corrosion current (I_{corr}), corrosion potential (E_{corr}), pitting corrosion potential (E_{pit}) and corro-sion rate (mm/year) were calculated from Tafel curve. The tests were carried out at room tempe-rature in 0.1 M NaCl solutions at a fixed and neutral pH value. The corroded samples were cleaned in distilled water and examined under optical light and scanning electron microscope.

Electrochemical impedance measurements: As in potentiodynamic polarization test, three electrode cell arrangements were also used in electrochemical impedance measurements. Rectangular samples (10 x 5 mm) were connected with copper wire and adopted as working electrode. EIS tests were performed in 0.1M NaCl solution at room temperature over a frequency range of 100 kHz to 0.2 Hz using a 5 mV amplitude sinusoidal voltage. The 10 x 5 mm sample surface was immersed in 0.1M NaCl solution (corrosion medium). All the measurements were performed at the open circuit potential (OCP). The test cells were maintained at room temperature and the NaCl solution was refreshed regularly during the whole test period. The impedance spectra were collected, fitting the experimental results to an equivalent circuit (EC) using the Echem Analyst TM data analysis software and evaluating the solution resistance (R_s), polarization resistance or charge transfer resistance (R_{ct}) and double layer capacitance (C_p) of the thermal treated alloys.

Results and discussion

Impedance measurements

Table 1 shows the electrochemical impedance spectroscopy (EIS) test results.

Table 1. *Impedance test results*

Alloy Code	Alloy Compositions	R_s / Ω	R_{ct} / kΩ	C_p / μF	OCP, V *vs.* SCE
Alloy-1	Al-6Si-0.5Mg	40.37	15.57	1.259	-0.8454
Alloy-2	Al-6Si-0.5Mg-0.5Cu	43.93	25.75	1.793	-0.7037
Alloy-3	Al-6Si-0.5Mg-1Cu	44.08	27.13	3.219	-0.6534

OCP versus time behavior

The open circuit potential (OCP) with exposure time of aged Al-6Si-0.5Mg-xCu alloys in 0.1 M NaCl solution is shown in Table1. Large fluctuations in open circuit potential for the alloys were seen during the time of 100 s exposure. After a period of exposure the OCP fluctuation decreased and reached steady state. The steady state OCP of Cu free alloy (Alloy-1) is -0.8454 V and it is the most negative OCP value among the alloys under investigation. The occurrence of a positive shift in OCP in the Al-6Si-0.5Mg alloys containing 0.5 and 1.0 % Cu indicates the existence of anodically controlled reaction. The OCP values mainly depend on the chemical compositions and thermal history of the alloys.

The data obtained were modeled and the equivalent circuit that best fitted to the experimental data is shown in Figure 1. R_s represent the ohmic solution resistance of the electrolyte. R_{ct} and C_p are the charge transfer resistance and electrical double layer capacitance respectively, which correspond to the Faradaic process at the alloy/media interface.

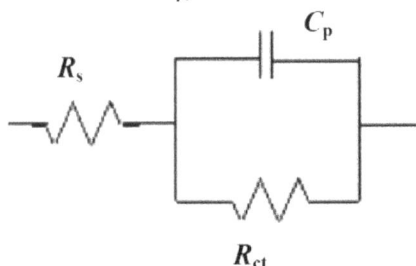

Figure 1. *Electrical equivalent circuit used for fitting of the impedance data of Al-6Si 0.5Mg-xCu alloys in 0.1M NaCl solution.*

Figure 2 shows the Nyquist diagrams (suggested equivalent circuit model shown in Figure 1) of the Al-6Si-0.5Mg-xCu alloys in 0.1M NaCl in de-mineralized (DM) water. In Nyquist diagrams, the imaginary component of the impedance (Z'') against real part (Z') is obtained in the form of capacitive-resistive semicircle for each sample.

Figure 3 shows the experimental EIS results in Bode magnitude diagram for Al-6Si-0.5Mg-xCu alloys. Bode plots show the total impedance behaviour against applied frequency. At high frequencies, only the very mobile ions in solution are excited so that the solution resistance (R_s) can be assessed. At lower intermediate frequencies, capacitive charging of the solid-liquid interface occurs. The capacitive value C_p can provide very important information about oxide properties when passivation or thicker oxides are formed on the surface. At low frequency, the capacitive charging disappears because the charge transfer of electrochemical reaction can occur and this measured value of the resistance corresponds directly to the corrosion rate. For this reason, this low frequency impedance value is referred to as polarization or charge transfer resistance (R_{ct}).

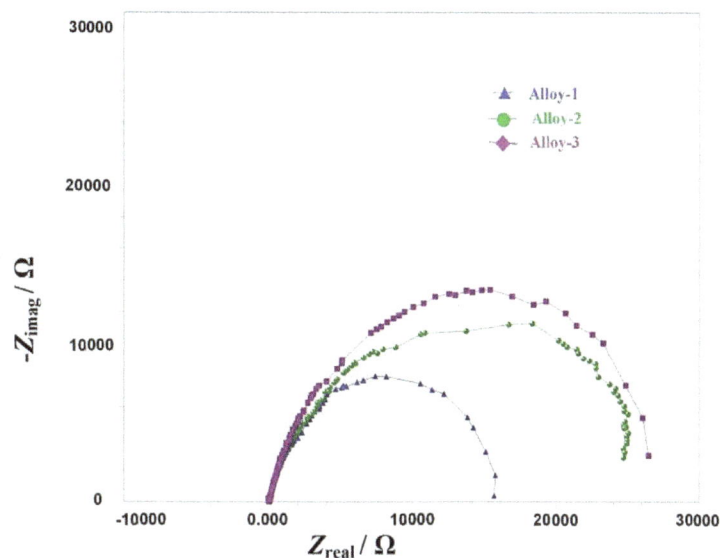

Figure 2. *Nyquist plots for the peak-aged Alloys 1, 2 and 3 in 0.1M NaCl solution.*

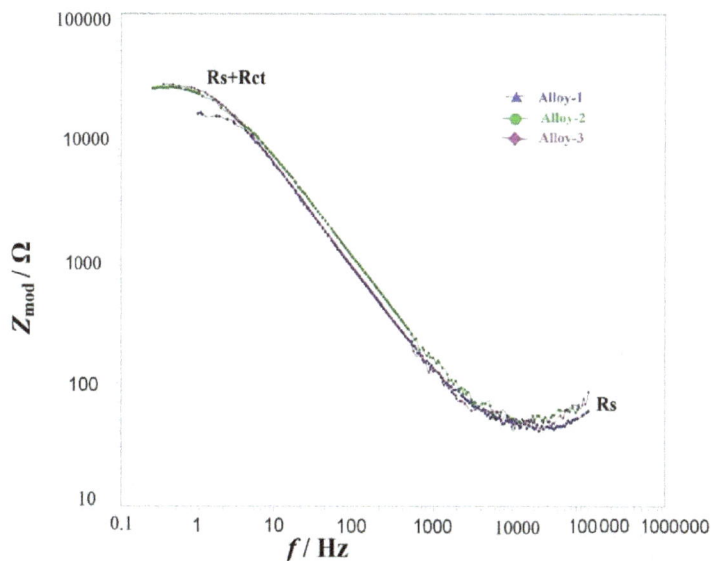

Figure 3. *Bode plots for the peak-aged Alloys 1, 2 and 3 in 0.1M NaCl solution.*

The solution resistance (R_s) of the alloys varies from 40-44 Ω (Table 1) and these values are very similar to each other. So there are insignificant changes of R_s values for the alloys during EIS testing. The R_s values are negligible with respect to R_{ct} and the electrolyte behaves as a good ionic conductor. Impedance measurements showed that in 0.1M NaCl solution, increasing Cu in the Al-6Si-0.5Mg alloys increases the charge transfer resistance (R_{ct}). For the Cu free Al-6Si-0.5Mg alloy (Alloy-1), the charge transfer resistance (R_{ct}) value in 0.1M NaCl solution is 15.57 kΩ, and this is increased to 25.75 and 27.13 kΩ with the addition of 0.5 and 1 wt % Cu to the Al-6Si-0.5Mg alloy respectively. The increase in the charge transfer resistance indicates an increase in the corrosion resistance of the alloys with Cu addition. The double layer capacitance (C_p) of the Cu free Al-6Si-0.5 Mg alloy (Alloy-1) is 1.259 μF, which is the lowest value among the alloys investigated. The double layer capacitance of Al-6Si-0.5Mg alloy increased with an increase in Cu content and the maximum was found for Alloy-3.

Potentiodynamic polarization measurements

Table 2 shows the potentiodynamic polarization test results obtained from the electrochemical tests.

Table 2. *Potentiodynamic polarization test results*

Alloy code	I_{corr} / μA	E_{corr} / mV	E_{pit} / mV	Corrosion rate, mm/year
Alloy-1	6.300	-764	-480	5.287
Alloy-2	5.640	-657	-408	4.732
Alloy-3	2.950	-697	-370	2.474

Potentiodynamic polarization curves of Al-6Si-0.5Mg-xCu alloys in 0.1M NaCl solution are shown in Figure 4. Anodic current density of Al-6Si-0.5Mg-xCu alloys decreased with Cu addition. This is caused by the slowing of the anodic reaction of Al-6Si-0.5Mg-xCu alloy.

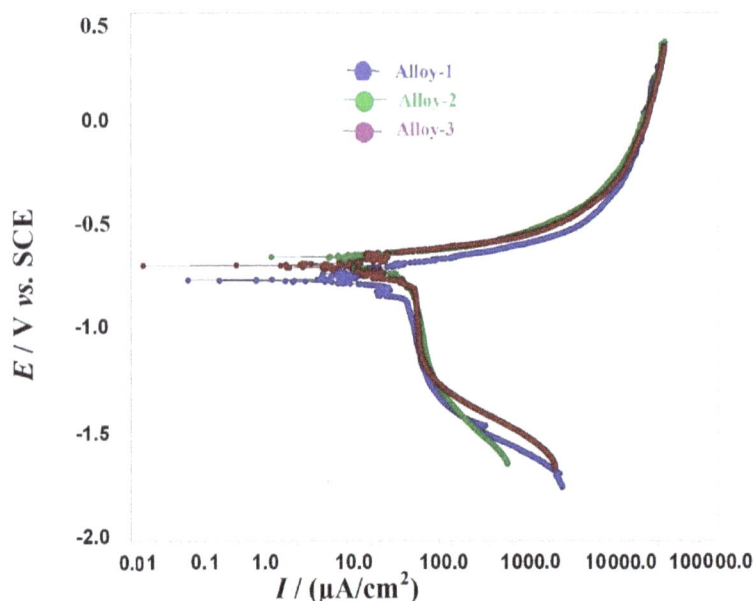

Figure 4. *Potentiodynamic polarization curves of peakaged Alloys 1, 2 and 3 in 0.1M NaCl solution.*

The addition of Cu caused the formation of micro-galvanic cells in α-aluminum matrix. The different intermetallic compounds (like Mg_2Si, Al_2Cu etc.) can lead to the formation of micro-galvanic cells because of the difference of corrosion potential between intermetallics and α-aluminum matrix. Park [13] has also reported that the addition of Cu increased the corrosion potential of a number of Al–Cu–Si alloys. For the Cu free Al-6Si-0.5Mg alloy (Alloy-1) corrosion potential is -764 mV, which is the highest negative potential among the alloys investigated. With increasing Cu, the corrosion potential of the alloys shifted towards more positive values. Pitting potential (Epit) of all Cu content alloys also shifted towards more positive values (from -480 mV to -370 mV). Potentiodynamic tests showed that in 0.1M NaCl solution, increasing Cu in the Al-6Si-0.5Mg alloy decreases the corrosion current (I_{corr}). For the Cu free Al-6Si-0.5Mg alloy (Alloy-1), the corrosion current (I_{corr}) value in 0.1M NaCl solution is 6.3 μA, and this decreased to 5.640 and 2.950 μA with the addition of 0.5 and 1 wt % Cu to the Al-6Si-0.5Mg alloy respectively and the corresponding corrosion rate decreases for these alloys (Alloy-2 = 4.732 mpy and Alloy-3 = 2.474 mm/year).

Microstructural Investigation

The microstructure of some selected as-corroded samples was observed under OLM and SEM. There was evidence of corrosion products of intermetallic compounds in all the samples examined. Besides, several pits were visible in all the samples examined. It is probable that the pits are formed by the intermetallics dropping out from the surface due to the dissolution of the surrounding matrix. However, it is also possible that the pits are caused by selective dissolution of the intermetallic/or particles of the second phase precipitates.

Osorio *et al.* [14] have demonstrated that in Al-Cu-Si alloys a more finely and homogeneously distributed Al_2Cu and needle-like Si particles in the ternary eutectic mixture, tend to improve the corrosion resistance mainly due to the galvanic protection of both Al_2Cu and Si phases [14]. Although it has also been reported [14,15] that fine Si particles tends to decrease the corrosion resistance of binary Al–Si alloys when associated with the Al_2Cu intermetallic phase, a better galvanic protection is provided for finer Al–Cu–Si alloy microstructures. It was also reported [16] that the ternary eutectic mixture consisting of Al + Al_2Cu + Si phases is nobler than the Al-matrix and Al-phase in the eutectic mixture [16].

Consequently, the forms of corrosion in the studied Al-6Si-0.5Mg-xCu alloys are slightly uniform and predominantly pitting corrosion as obtained by the OLM and SEM. Samples were characterized by OLM and SEM following potentiodynamic polarization tests. The peakaged Cu free alloy (Alloy-1) exhibited pits on their surface (Figure 5), which apparently had nucleated randomly.

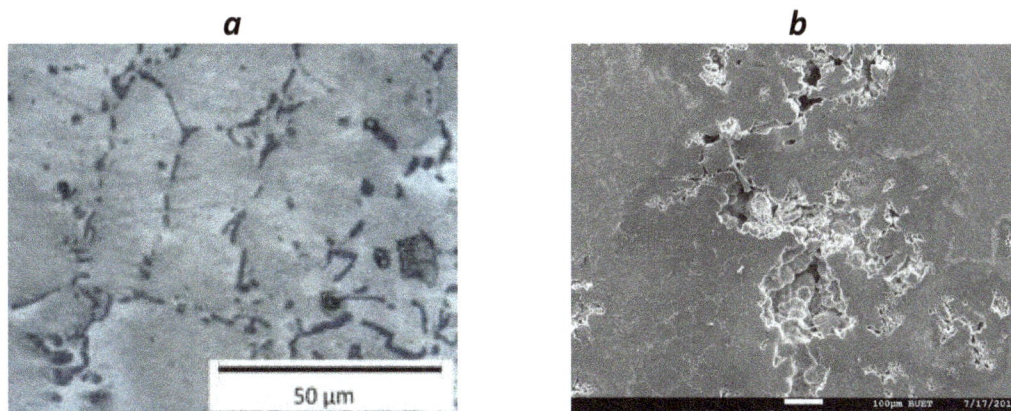

Figure 5. (a)OLM and **(b)** SEM images show the damage surface morphology of as-corroded T6 aged Alloy-1 in 0.1M NaCl solution.

Conversely, the exposed surface of the alloys exhibited a corrosion product with a rippled appearance covering the surface after polarization. All the optical micrographs (Figures 5-6) also showed that there was no corrosion in the fragmented and modified Al-Si eutectics.

a   *b*

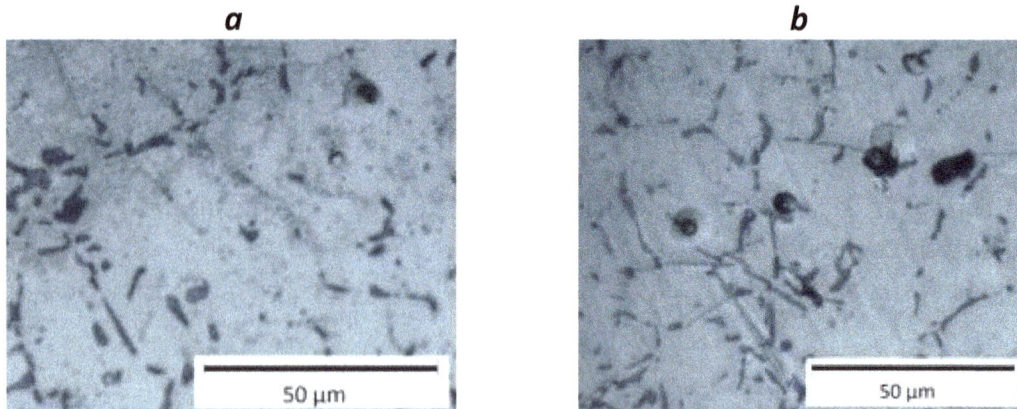

Figure 6. OLM images of the damage surfaces morphology of as-corroded T6 aged (a) Alloy-2 and (b) Alloy-3 in 0.1M NaCl solution.

Conclusions

The following conclusions may be drawn from the above investigation:

1. The EIS tests have shown that the additions of Cu into Al-Si-Mg alloy tend to increase the excellent corrosion resistance of Al-Si-Mg alloy in NaCl media. The corrosion resistance, Rct value of the alloys shows a maximum at 1 wt % Cu.

2. The linear polarization and Tafel extrapolation plot show that the corrosion current (I_{corr}) and corrosion rate (mm/year) decrease with increasing of Cu content into Al-6Si-0.5Mg alloy. The open circuit potential (OCP), corrosion potential (E_{corr}) and pitting corrosion potential (E_{pit}) in the NaCl solution were shifted in the more noble direction due to Cu additions into Al-6Si-0.5Mg alloy.

3. Consequently, the forms of corrosion in the studied Al-6Si-0.5Mg-xCu alloys are pitting corrosion as obtained from the microstructures study with pits observations.

References

[1]   M. G. Fontana, N.D. Greene, *Corrosion Engineering,* McGraw-Hill book Company, New York, 1987, 8-29.

[2]   S. Zor, M. Zeren, H. Ozkazance, E. Karakulak, *Anti-Corrosion Methods and Materials* **57** (2010) 185-191.

[3]   G. M. Scamans, J. A. Hunter, N. J. H. Holroyd, , *Proc. of 8th Inter. Light metals Congress,* Leoban Wien, 1989, 699-705,

[4]   M. Czechowski, *Adv.* Mater *Sci.* **7** (2007) 13-20.

[5]   M. Abdulwahab, I.A. Madugu, S.A. Yaro, A.P.I. Popoola, *Journal of Minerals & Materials Characterization & Engineering* **10** (2011) 535-551.

[6]   G. Svenningsen, M.H. Larsen, *Corros. Sci.* **48** (2006) 3969–3987.

[7]   G. Svenningsen, J.E. Lein, A. Bjorgum, J.H. Nordlien, K. Nisancioglu, *Corros. Sci.* **48** (2006) 226-242.

[8]   M. H. Larsen, J. C. Walmsley, *Mater. Sci. Forum* **519-521** (2006) 667-671.

[9]   G. Svenningsen, M.H. Larsen, J.H. Nordlien, K. Nisancioglu, *Corros. Sci.* **48** (2006) 258–272.

[10] G. Svenningsen, M.H. Larsen, *Corros. Sci.* **48** (2006) 1528–1543.

[11]   H. Zhan, J. M. C. Mo, F. Hannour, L. Zhuang, H. Terryn, J. H. W. de Wit, *Materials and Corrosion* **59** (2008) 670–675.

[12]   M. H. Larsen, J. C. Walmsley, O. Lunder and K. Nisancioglu, *J. Electrochem. Soc.* **157** (2010) 61-68.

[13]   M. Park, *J. Mater. Sci.* **40** (2005) 3945.

[14]   W. R. Osório, D. J. Moutinho, L. C. Peixoto, I. L. Ferreira, A. Garcia. *Electrochimica Acta* **56** (2011) 8412–8421.

[15]   W. R. Osório, N. Cheung, J. E. Spinelli, A. Garcia. *J. Solid State Electrochem* **11(10)** (2007) 1421-1427.

[16]   W. R. Osório, L. C. Peixoto, D. J. Moutinho, L. G. Gomes, I. L. Ferreira, A. Garcia. *Materials and Design* **32** (2011) 3832–3837.

16

Corrosion inhibition of carbon steel by extract of *Buddleia perfoliata*

ROY LOPES-SESENES, JOSE GONZALO GONZALEZ-RODRIGUEZ$^{\boxtimes}$, GLORIA FRANCISCA DOMINGUEZ-PATIÑO*, ALBERTO MARTINEZ-VILLAFAÑE**

Universidad Autonoma del Estado de Morelos, CIICAP, Av. Universidad 1001, 62209-Cuernavaca, Mor.,Mexico
*Universidad Autonoma del Estado de Morelos, Facultad de Ciencias Biologicas, Av. Universidad 1001, 62209-Cuernavaca, Mor., Mexico
**Centro de Investigaciones en Materiales Avanzados, Miguel Cervantes 120, Chihuahua, Mexico

$^{\boxtimes}$Corresponding Author

Abstract

Buddleia perfoliata *leaves extract has been investigated as a carbon steel corrosion inhibitor in 0.5 M sulfuric acid by using polarization curves, electrochemical impedance spectroscopy and weight-loss tests at different concentrations (0, 100, 200, 300, 400 and 500 ppm) and temperatures, namely 25, 40 and 60 °C. Results showthat inhibition efficiency increases as the inhibitor concentration increases, decreases with temperature, and reaches a maximum value after 12 h of exposure, decreasing with a further increase in the exposure time. It was found that the inhibitory effect is due to the presence of tannines on this extract.*

Keywords

Corrosion inhibitor, *Buddleia perfoliata*, EIS, polarization curves.

Introduction

Due to currently imposed requeriments for eco-friendly corrosion inhibitors, there is a growing interest in the use of natural products such as leaves or seeds extracts. Some papers have reported the use of natural products for mild steel corrosion inhibition in different environments [1-23]. This is due to the fact that sinthetic inhibitors are, among other factors, expensive and highly toxic. Among the so-called "green inhibitors" there are organic compounds, such as ascorbic acid, succinic acid, tryptamine, caffeine, *etc.,* that act by adsorption on metal surface. Additionally, some other natural products such as black pepper, *Azadirachta indica, Gossipium hirsutum,* guanadine, *Occimum viridis, Talferia occidentalis* and *Hibiscus sabdariffa*, were used. For instance, Oguzie carried out the inhibitive action of leaf extracts of *Sansevieria trifasciata* on aluminium corrosion in 2 M HCl and 2 M KOH solutions by using the gasometric technique [20]. Results indicated that the extract functioned as a good inhibitor in both environments and inhibition efficiency increased with the increase of the inhibitor concentration. A mechanism of physical adsorption is proposed for the inhibition behaviour. The adsorption characteristics of the inhibitor were approximated by Freundlich isotherm. In another work by Chauhan *et al.* [3] the inhibition effect of *Zenthoxylum alatum* plant extract on the corrosion of mild steel in 5 % and 15 % aqueous hydrochloric acid solution has been investigated by weight-loss method and electrochemical impedance spectroscopy (EIS). The corrosion inhibition efficiency increased by increasing the plant extract concentration till 2400 ppm. The adsorption of this plant extract on the mild steel surface obeys the Langmuir adsorption isotherm. Okafor *et. al.* [4] studied the inhibitive action of leaves, seeds and a combination of leaves and seeds extracts of *Phyllanthus amarus* on mild steel corrosion in HCl and H_2SO_4 solutions using weight- loss and gasometric techniques. The results indicate that the extracts functioned as a good inhibitor in both environments, and inhibition efficiency increases with extracts concentration. The adsorption characteristics of the inhibitor were approximated by Temkin isotherm. The corrosion efficiency of these extracts is normally ascribed to the presenceof complex organic species such as tannins, alkaloids and nitrogen bases, carbohydrates and proteins as well as their acid hydrolysis products.

The genus *Buddleia*, included in the family *Loganiaceae*, and previously classified in a family of its own, the *Buddlejaceae*, is now classified in the family *Scrophulariaceae*. Native to Asis, Africa, North and South America, *Buddleia* is a genus containing 100 species, 50 being distributed in America, of which 16 grow in Mexico [22]. *Buddleia* species are widespread and share some remarkable similarities in their medicinal uses. This may well indicate the presence of the same or similar compounds with a particular pharmacological action. A patterns is emerging about the composition of these compounds; flavonoid and iridoid glycosids being the major seccondary metabolites that have been isolated to date [23]. *Buddleia perfoliata* became officialy recognized in the 1930 Mexican Pharmacopoeia where it was shown to have antisudorific activity [24]. This plant also contains essential oil, tannic, gallic and oxalic acids [25]. In folk medicines, it is used in the treatment for tuberculosis as well as for catarrh, ptyalism and headaches [25]. In the present paper, we evaluated the inhibitory effect of *Buddleia perfoliata* in the corrosion of 1018 carbon steel in 0.5M H_2SO_4 by using both gravimetric and electrochemical techniques.

Experimental procedure

Corrosion tests were performed on coupons prepared from 1018 carbon steel rods containg 0.14 % C, 0.90 % Mn, 0.30 % S, 0.030 % P and as balance Fe, encapsulated in commercial epoxic resin with the exposed area of 1.0 cm². The aggressive solution, 0.5 M H_2SO_4 was prepared by dilution of analytical grade H_2SO_4 with double distilled water. Dried *Buddleia perfoliata* leaves (38.7 g) were soaked in 300 ml of methanol during 24 h and refluxed during 5 h obtaining a solid, which was weighted and dissolved in methanol untill this was completely evaporated and used as a stock solution and then for preparation of the desired concentrations by dilution. Methanol is commonly used in the obtaining green inhibitors [2-11] and since it completely evaporates, there is no risk of toxicity. Weight-loss experiments were carried out with steel rods of 2.5 cm in length and 0.6 cm diameter abraded with fine 1200 grade emery paper, rinsed with acetone, and exposed to the aggressive solution during 72 h. After a total exposition time of 72 h, specimens were taken out, washed with distilled water, degreased with acetone, dried and weighed accurately. Specimens were hanging by nylon fibers which were introduced through a hole of 0.5 mm in diameter, drilled in one end of the specimen. Tests were performed by triplicate at room temperature (25 °C), 40 and 60 °C by using a hot plate. Corrosion rates, in terms of weight loss measurements, ΔW, were calculated as follows:

$$\Delta W = (m_1 - m_2) / A \tag{1}$$

were m_1 is the mass of the specimen before corrosion, m_2 the mass of the specimen after corrosion, and A the exposed area of the specimen. For the weight loss tests, inhibitor efficiency, IE, was calculated as follows:

$$IE = 100 \, (\Delta W_1 - \Delta W_2)/\Delta W_1 \tag{2}$$

where ΔW_1 is the weight loss without inhibitor, and ΔW_2 the weight loss with inhibitor. Specimens were removed, rinsed in water and in acetone, dried in warm air and stored in desiccators. Specimens were weighed in an analytical balance with a precision of 0.1 mg. Surface analysis of corroded specimens was carried out by a Scanning Electronic Microscope (SEM). Electrochemical techniques employed included potentiodynamic polarization curves and electrochemical impedance spectroscopy (EIS) measurements. In all the experiments, carbon steel electrode was allowed to reach stable open circuit potential value, E_{corr}. Each polarization curve was recorded three times at constant sweep rate of 1 mV s^{-1} at the interval from -1000 to + 1500 mV in respect to the E_{corr} value. Measurements were obtained by using a conventional three electrodes glass cell with two graphite electrodes symmetrically situated, and a saturated calomel electrode (SCE) as a reference electrode with a Lugging capillary bridge. Corrosion current density values, i_{corr}, were obtained by using Tafel extrapolation. Inhibitor efficiency, IE, was calculated as follows:

$$IE = 100 \, (i_{corr1} - i_{corr2})/i_{corr1} \tag{3}$$

where i_{corr1} is the corrosion current density value without inhibitor, and i_{corr2} the corrosion current density value with inhibitor. EIS tests were carried out three times at E_{corr} by using a signal with amplitude of 10 mV in a frequency interval of 100 mHz - 100 KHz. An ACM potentiostat, controlled by a desk top computer was used for the polarization curves, whereas for the EIS measurements, a model PC4 300 Gamry potentiostat was used.

Results and discussion

The effect of *Buddleia perfoliata* concentration in the polarization curves at 25 °C is shown in Fig. 1, whereas electrochemical parameters for these curves are shown in Table 1. It can be seen that steel exhibits an active-passive behavior with and without inhibitor; in the uninhibited solution, the steel shows an increase in the anodic current density, but around 400 mV a region where the current remains more or less stable, similar to a passive region, apperas although very unstable since some anodic transients can be seen due to the brakedown and repair of this incipient passive layer. This unstable region dissapears with the addition of 100, 200 or 300 ppm of *Buddleia perfoliata* and the passive zone becomes stable, but with a further increase of the inhibitor concentration this unstable regions apperas once again. It is clear that the addition of *Buddleia perfoliata* has caused a clear decrease in both the anodic and cathodic branch of the polarization curves, and this effect is more pronounced as the inhibitor concentration increases; from Table 1 it can be seen that as soon as the extract is added to the electrolyte, the E_{corr} value becomes more negative and the i_{corr} value decreases, reaching its lowest value when 500 ppm of inhibitor is added.

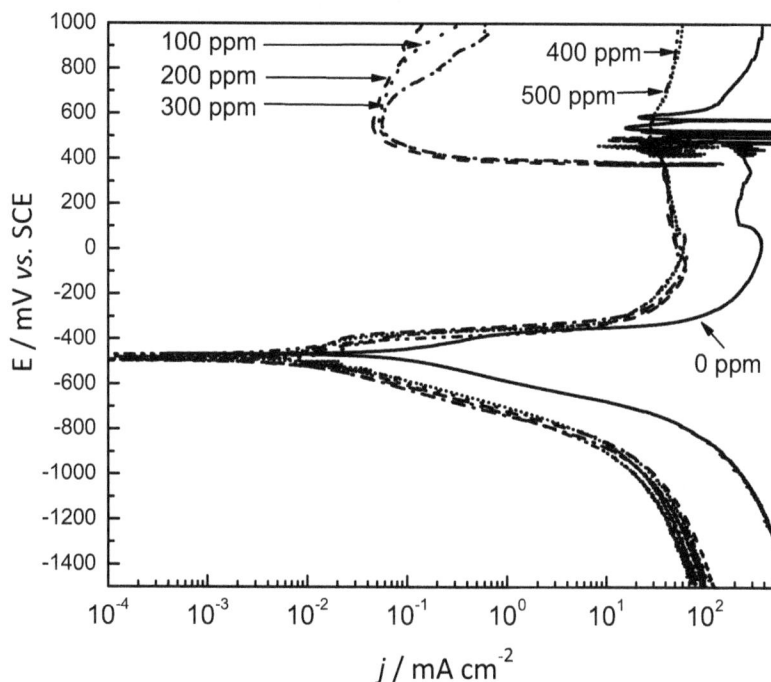

Figure 1. *Effect of Buddleia perfoliata concentration in the polarization curves for 1018 carbon steel corroded in 0.5 M H$_2$SO$_4$ at 25 °C*

Table 1.*Electrochemical parameters obtained from polarization curves at 25 °C. Estimated error in potential is ± 6 mV, in current density is ± 5 x 10^{-3} mA cm^{-2} and in resistance is ± 1 x 10^{-3} Ω cm^2.*

c_{inh} ppm	E_{corr} mV *vs.* SCE	i_{corr} mA cm^{-2}	i_{pas} mA cm^{-2}	β_a mV dec^{-1}	β_c mV dec^{-1}	R_p Ω cm^2
0	-473	0.09	50	125	-120	60
100	-503	0.07	0.06	118	-115	102
200	-497	0.07	0.05	108	-115	145
300	-491	0.06	0.04	65	-110	240
400	-460	0.05	25	60	-105	565
500	-478	0.05	27	50	-105	600

As expected, the polarization resistance value, R_p, increases as the inhibitor concentration increases, reaching a maximum value when 500 ppm of inhibitor is added. In addition to this, passive current density decreases as the *Buddleia perfoliata* concentration increases up to 300 ppm, but increasing again with a further increase of the inhibitor concentration. The cathodic slope was practically unaffected by the addition of the *Buddleia perfoliata*, which indicates that hydrogen evolution reaction is diminished exclusively by the surface blocking effect of adsorbed inhibitor [15]. Regarding the anodic region of the potentiodynamic polarization curves, there is clearly an active-passive behavior either in presence or in absence of the inhibitor. Also, the currents remains almost the same in all cases in the active dissolution region of the metal, but it decreased in the passive region when the inhibitor is added. This behavior could be related to a change in the anodic reaction mechanism (iron dissolution) which is corroborated by a decrease in the anodic Tafel slope with increasing concentration of *Buddleia perfoliata* [15].

Nyquist and Bode diagrams for 1018 carbon steel exposed to 0.5 M H_2SO_4 with different *Buddleia perfoliata* dosis is shown in Fig. 2. Nyquist diagram, Fig. 2a, shows a single deppressed, capacitive semicircle with its center in the real axis regardless of the inhibitor concentration, indicating that the corrosion process is under charge transfer control from the metal to the electrolyte through the electrochemical double layer. This type of plot is characteristic of solid electrodes and it is often ascribed to dispersion effects, which are attributed to roughness and inhomogeneities of the surface during corrosion [24,25].

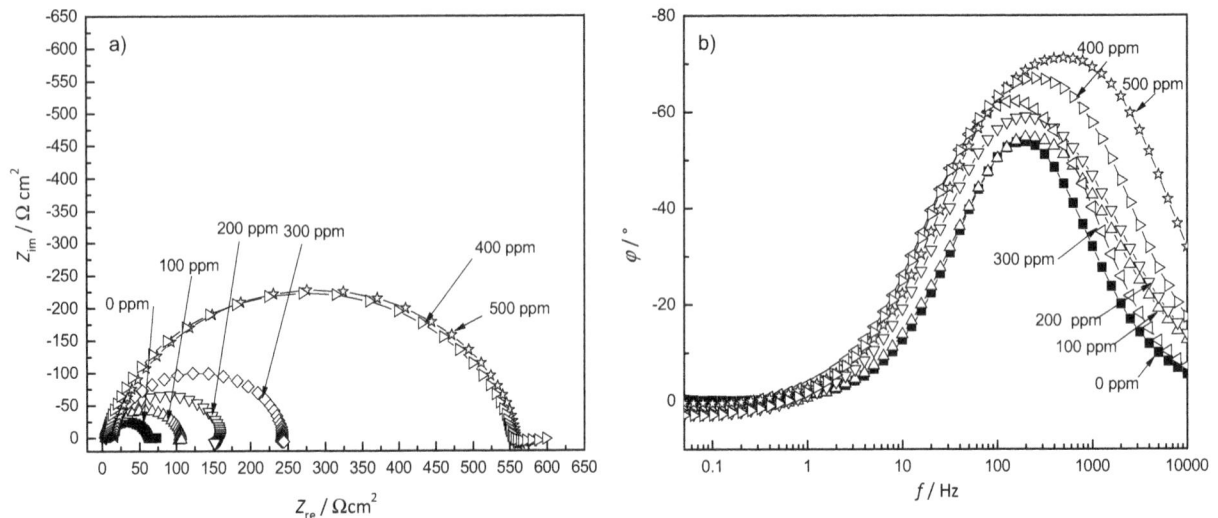

Figure 2. *Effect of Buddleia perfoliata concentration in the a) Nyquist and b) Bode diagrams for 1018 carbon steel corroded in 0.5 M H_2SO_4 at 25 °C*

This behavior is not affected by the presence of the inhibitor, indicating the activation-controlled nature of the reaction; the semicircle diameter increases with the inhibitor concentration, reaching a maximum value with 500 ppm of inhibitor. These results support those obtained from the polarization curves and confirm the inhibitor adsorption onto carbon steel surface. The intersection of of the semicircle with the real axis at high frequencies provides a value of the solution resistance of 4.3 Ω cm^2. The semircle diameter is related to the charge transfer resistance, R_{ct}, inversely proportional to the i_{corr} value, thus, the lowest corrosion rate is attained with 500 ppm, as indicated by the polarization curves in Fig. 1 and Table 1. For the uninhibited solution, an R_{ct} value of 52 Ω cm^2 was found. Bode diagram, Fig. 2b, shows a single peak around 200 Hz, which shifts towards higher frequency values as the inhibitor concentration increases up to 500 ppm.

Equivalent electric circuit used to simulate the EIS data for 1018 carbon steel exposed to 0.5 M H_2SO_4 with different *Buddleia perfoliata* dosis is shown in Fig. 3. In Fig. 3, R_s represents the solution resistance, R_{ct} the charge transfer resistance or the resistance to the flow of electrons from the metal to the electrolyte, and C_{dl} the capacitance of the electrochemical double layer, or the capacity to storage charge in this layer. However, one has to account for the inhomogeneity of the surface-electrolyte system. When a non-ideal frequency response is present, it is commonly accepted to employ distributed circuit elements in an equivalent circuit. The most widely used is a constant phase element (CPE) or time constant, which has a non-integer power dependence on the frequency. Often a CPE is used in a model in place of a capacitor to compensate for non-homogeneity in the system. Since only a single peak exists in Bode diagram, Fig. 2b, only one time constant is needed to simulate the EIS data. Table 3 summarizes the calculated parameters to simulate the EIS data using circuit shown in Fig. 3. It can be seen that the R_{ct} value reaches its highest value between 400 and 500 ppm, whereas the electrochemical double layer capacitance, C_{dl}, attains its lowest value at these inhibitor concentrations, which may be attributed to the adsorption of the components in the *Buddleia perfoliata* extract onto the metal/electrolyte interface or to an increase of the double layer [12].

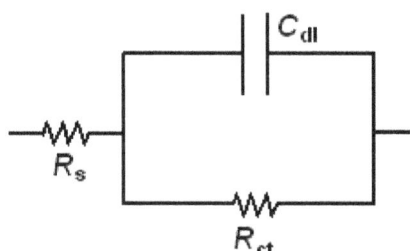

Figure 3. *Electric circuit used to simulate the EIS data*

Table 3. *Calculated parameter values used to simulate the impedance data for EIS data Estimated error in resistance is ± 1 x 10^{-3} Ω cm^2 whereas for capacitance it is ± 1 x 10^{-8} F cm^{-2}*

c_{inh} / ppm	R_s / Ω cm^2	R_{ct} / Ω cm^2	C_{dl}/ F cm^{-2}
0	4.9	52	5.3 x 10^{-5}
100	6.9	102	7.8 x 10^{-6}
200	5.2	145	5.3 x 10^{-6}
300	4.9	240	4.1 x 10^{-6}
400	4.4	565	1.7 x 10^{-6}
500	4.2	600	4.7 x 10^{-7}

Alternatively, the double layer capacitance, C_{dl}, was calculated from the equation bellow:

$$C_{dl} = 1/2\pi f_{max} R_{ct} \tag{4}$$

where f_{max} is the frequency value at which the imaginary component of the impedance is maximal. For the uninhibited solution, a C_{dl} value of 530 μF cm^{-2} was found. Table 3 gives the results for the R_s, R_{ct} and C_{dl} values for the solution with and without inhibitor. It is important to note that theincrease of the inhibitor concentration brings an increase in the charge transfer resistance value and a decrease in the double layer capacitance. The decrease in the C_{dl} value can be interpreted as due to the adsorption of the inhibitor onto the electrode surface [22]. The double layer formed at the metal-solution interface is considered as an electric capacitor, whose capacitance decreases due to the displacement of water molecules and other ions originally adsorbed on the electrode by the inhibitor molecules, forming a protective film. The thickness of the film formed increases with increasing concentration of the inhibitor, since more inhibitor adsorbs on the surface, resulting in lower C_{dl} values.

The stability of any film formed by the inhibitor was evaluated by plotting the Nyquist diagram at different times during 24 h, as shown in Fig. 4. This figure shows that the semicircle diameter remains more or less constant during 8 h, and after this time, the semicircle diameter decreases continuously as time elapses. The change in the R_{ct} and C_{dl} values calculated from this figure are shown in Fig. 5, where it is evident that the R_{ct} values remains more or less constant during the first 8 h, but after this, its value starts to decrease, indicating that the inhibitor remains more or less stable on the steel surface during this time. The C_{dl} value increases as time elapses, indicating a decrease in the inhibitor film thickness.

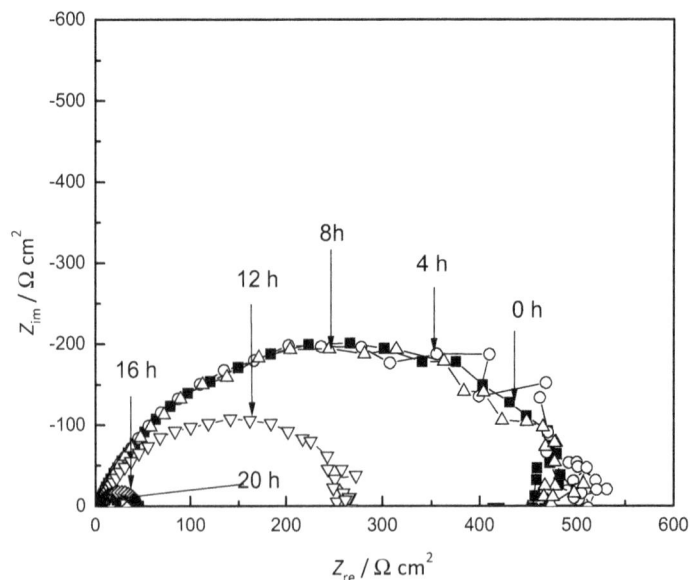

Figure 4. *Change of the Nyquist diagram with time for 1018 carbon steel corroded in 0.5 M H₂SO₄ at 25 °C with the addition of 500 ppm of Buddlia perfoliata*

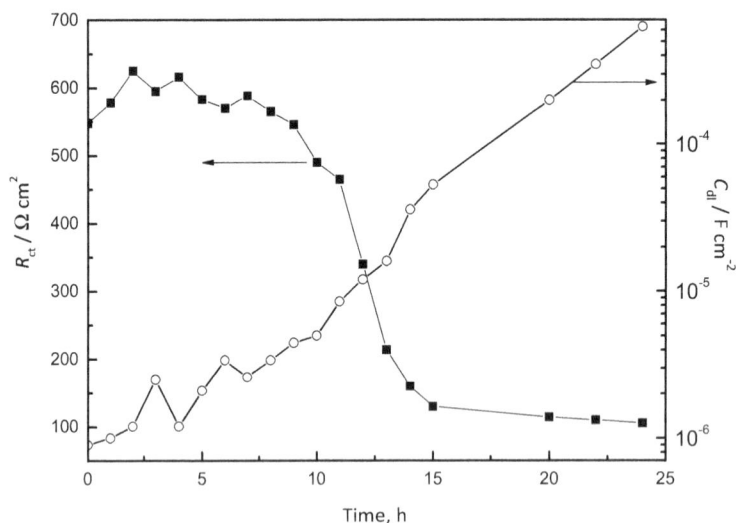

Figure 5. *Change of the charge transfer resistance, R_{ct}, and double layer capacitance values, C_{dl}, with time for 1018 carbon steel corroded in 0.5 M H₂SO₄ at 25 °C with the addition of 500 ppm of Buddleia perfoliata*

The R_{ct} values were used to calculate the *IE* according to the equation:

$$IE = 100 \, (R_{ct2} - R_{ct1}) / R_{ct2} \tag{5}$$

where R_{ct1} and R_{ct2} are the charge transfer resistance values for the uninhibited and inhibited solution, respectively. The results, together with those obtained by polarization curves (Tafel method), weight-loss and polarization resistance measurements are given in Table 2. It can be clearly seen that all different techniques show that the corrosion inhibition efficiency increases with the increase of inhibitor concentration, reaching a maximum value at 500 ppm of inhibitor. The discrepancy in the *IE* values obtained from different techniques can be interpreted as the result of different measurements time. Therefore, these results suggest, once again, the formation of an insoluble inhibitor film due to the adsorption of inhibitor onto carbon steel surface.

Table 2. *Efficiency of Buddleia perfoliata as an inhibitor of 1018 carbon steel in 0.5 M H_2SO_4 calculated with different techniques. Estimated error in efficieny values is ± 3 %.*

c_{inh} / ppm	$IE_{Tafel\ slope}$, %	$IE_{Weight-loss}$, %	IE_{EIS}, %	$IE_{Pol.\ resistance}$, %
100	22	10	47	41
200	22	15	65	58
300	33	18	75	69
400	44	22	84	80
500	44	27	88	84

In order to evaluate the adsorption process of *Buddleia perfoliata* on the 1018 carbon steel surface, Langmuir, Temkin and Frumkin adsorption isotherms were obtained according to the following equations:

Langmuir: $\theta/1-\theta = Kc_{inh}$ (6)

Temkin: $\log(\theta/c_{inh}) = \log K - g\theta$ (7)

Frumkin: $\log(\theta c_{inh})/(1-\theta) = \log K + g\theta$ (8)

where θ is the surface coverage, K is the adsorption-desorption equilibrium constant, c_{inh} is the inhibitor concentration and g is the adsorbate interaction parameter. The three isotherms tested fitted well the data obtained, as can be seen in the Fig. 6 indicating that *Buddleia perfoliata* is adsorbed onto the carbon steel surface. However, the isotherm which gave the best R^2 value, 0.996, was the Frumkin one. From the Frumkin isotherm, the adsorption-desorption equilibrium constant K was determined as 3.785 L mg^{-1} leading to an adsorption free-energy value of -37.4 kJ mol^{-1}. Generally, values of the adsorption free-energy much less than -40 kJ mol^{-1} have typically been correlated with the electrostatic interactions between organic molecules and charged metal surface (physisorption) whilst those values in the order of -40 kJ mol^{-1} are associated with charge sharing, or charge transfer from the organic molecules to the metal surface (chemisorption) to form a co-ordinate type of bond [26].The negative value of the free-energy of adsorption value means that the adsorption process is spontaneous, while the value around -40 kJ mol^{-1} indicates that *Buddleia perfoliata* was chemisorbed on steel surface. The Temkin isotherm, Fig. 5 b, also shows a good correlation with the experimental data, and the negative value of the slope indicates the existance of a repulsive lateral interaction in the adsorption layer [26].

The effect of temperature on the corrosion of carbon steel in the uninhibited and inhibited 0.5 M H_2SO_4 solutions was studied using both, potentiodynamic polarization curves and EIS tests. Fig. 7 shows the effect of temperature on the polarization curves for uninhibited and inhibited solution containing 500 ppm of inhibitor, respectively. It was found that the corrosion rate of steel in both, uninhibited and inhibited, solutions increases as the temperature increases. However, the extent of the rate increment in the inhibited solution is higher in the uninhibited than in the inhibited solution. This suggests that the corrosion inhibition might be caused by the inhibitor adsorption onto the steel surface from the acidic solution, and higher temperatures might cause a stronger adsorption of the inhibitor on the steel surface.

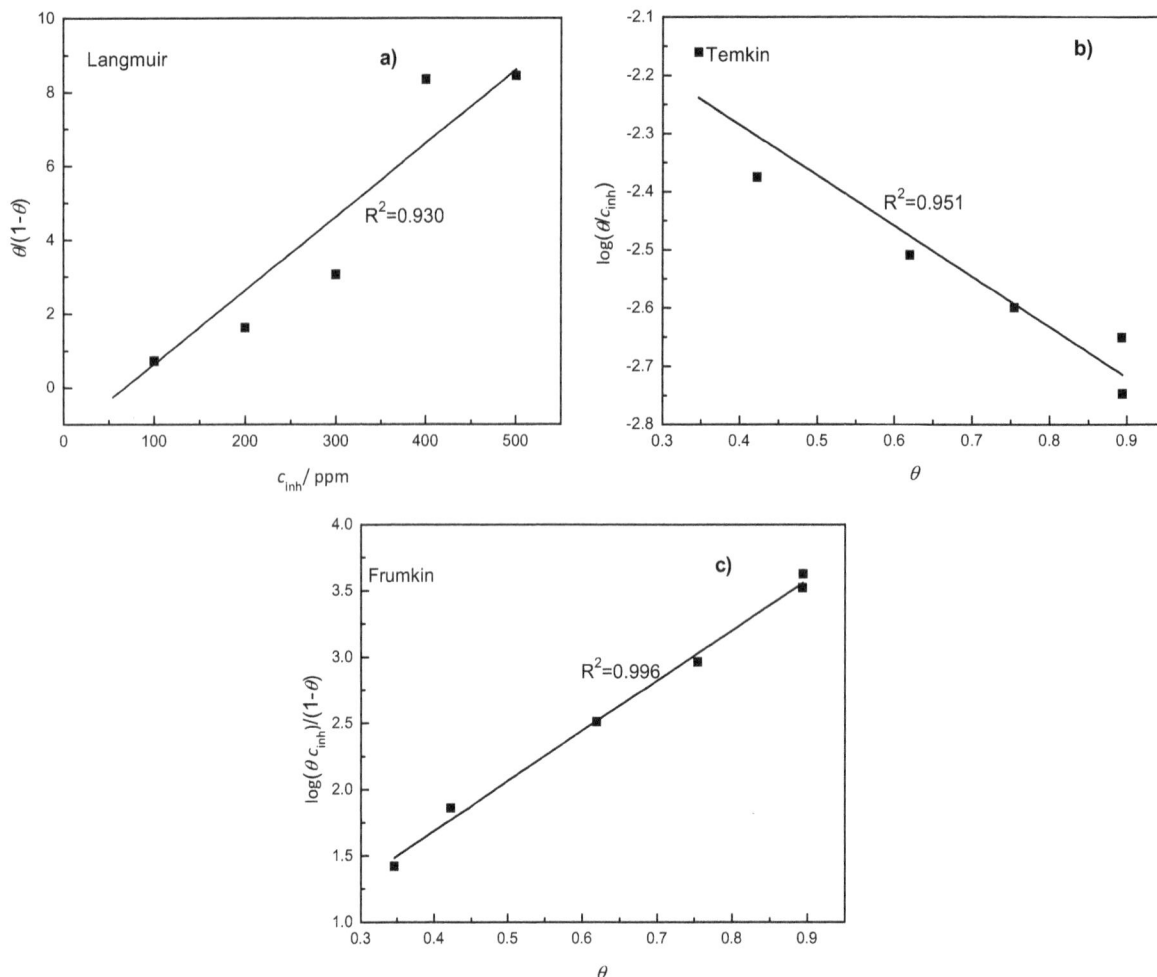

Figure 6. *a) Langmuir b) Temkin and c) Frumkin adsorption isotherm for 1018 carbon steel in 0.5 M H$_2$SO$_4$ at 25 °C with the addition of Buddleia perfoliata*

The apparent activation energy, E_a, associated with 1018 carbon steel in uninhibited and inhibited acid solution was determined by using an Arrhenius-type plot according to the following equation:

$$\log i_{corr} = -E_a / 2.303RT + \log F \tag{9}$$

where i_{corr} is the corrosion current density value, R is the molar gas constant, T is the absolute temperature and F is the frequency factor. Arrhenius plots of log i_{corr} against T^1 for 1018 carbon steel in 0.5 M H$_2$SO$_4$ in absence and presence of *Buddleia perfoliata* are shown in Fig. 8. The apparent activation energy obtained for the corrosion process in the inhibitor-free, uninhibited acid solution was found to be 83.9, and 63.9 kJ mol^{-1} in the presence of the inhibitor. Notably, the energy barrier of the corrosion reaction decreased in the presence of the inhibitor, which can be due to the chemisorption of the inhibitor on the steel surface.

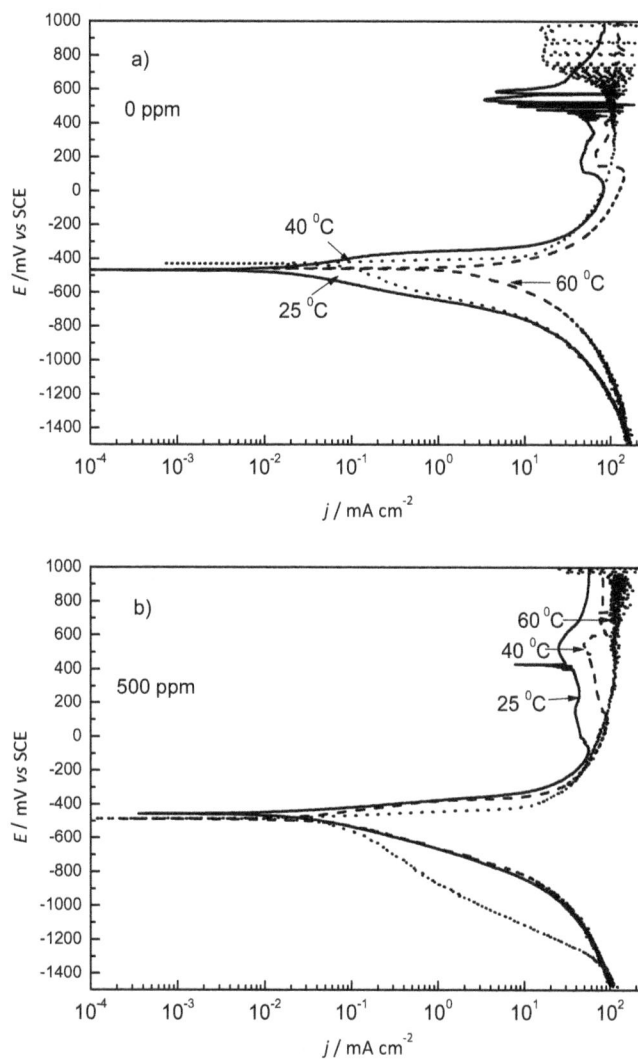

Figure 7. *Effect of temperature in the polarization curves for 1018 carbon steel corroded in 0.5 M H_2SO_4 a) wihout and b) with 500 ppm of Buddleia perfoliata*

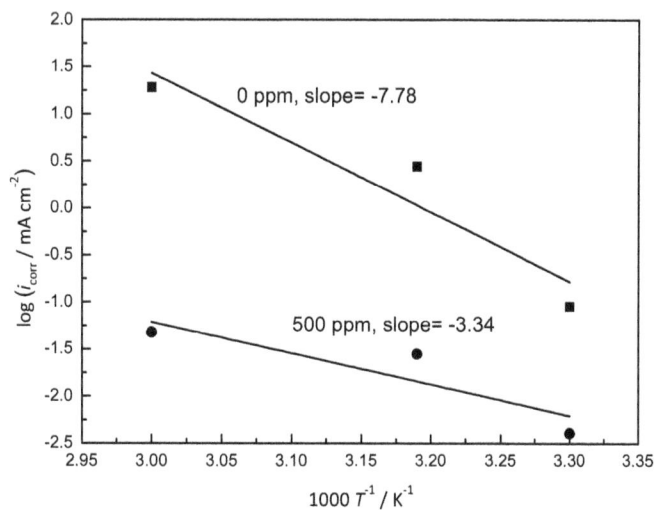

Figure 8. *Arrhernius plots for log (i_{corr}) vs. 1000T^{-1} for 1018 carbon steel corroded in 0.5 M H_2SO_4 wihout and with 500 ppm of Buddleia perfoliata*

In a way similar to the polarization curves, the effect of temperature on the Nyquist diagrams for uninhibited and inhibited solutions are shown in Figs. 9 and 10, respectively. In both cases it can be seen that the semicircle diameters, and thus, the R_{ct} values, decreased as the temperature increased, which was more evident for the inhibited solution. An Arrhernius plot of log $1/R_{ct}$ against T^{-1} for 1018 carbon steel in 0.5 M H_2SO_4 in absence and presence of *Buddleia perfoliata* is shown in Fig. 10, where apparent activation energy obtained for the corrosion process in the free acid solution was found to be 4.31 and 3.5 kJ mol^{-1} in presence of the inhibitor.

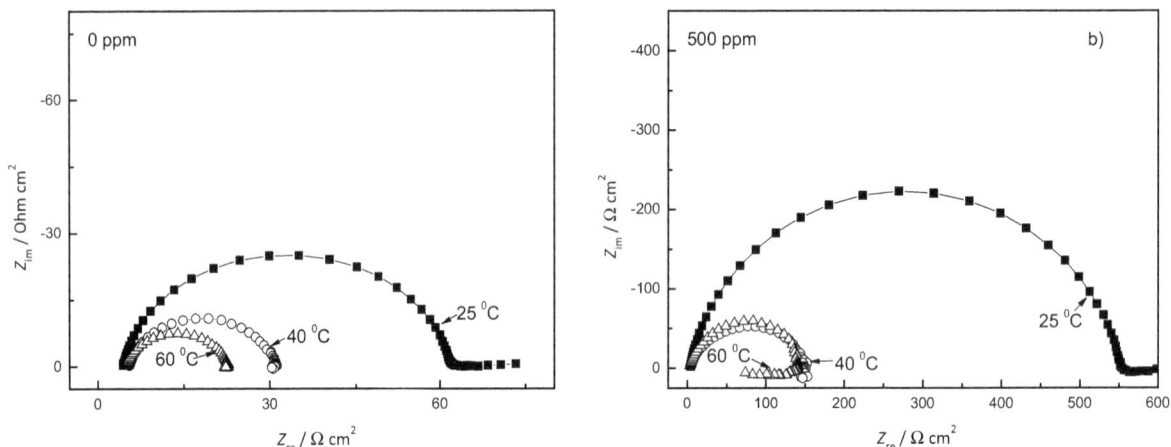

Figure 9. *Effect of temperature in the Nyquist diagrams for 1018 carbon steel corroded in 0.5 M H$_2$SO$_4$ a) wihout and b) with 500 ppm of Buddleia perfoliata*

According to Popova *et al.* [27] lower E_a values in solutions in presence of *Buddleia perfoliata* indicate a specific type of adsorption of the inhibitor, while Szauer and Brandt [28] associate this behavior with the chemisorption of the inhibitor to the metal surface. Taking into consideration these references and the E_a value calculated from the Arrhenius plots, the action of *Buddleia perfoliata* as a corrosion inhibitor for 1018 carbon steel in acid solution can be attributed to a strong type of chemisorption of the inhibitor onto metal surface.

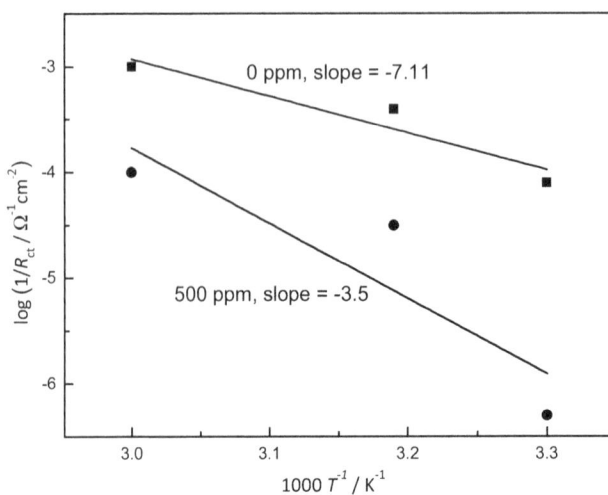

Figure 10. Arrhernius plots for log (1/R$_{ct}$) vs. 1000 T^{-1} for 1018 carbon steel corroded in 0.5 M H$_2$SO$_4$ without and with 500 ppm of Buddleia perfoliata

Some micrographs of 1018 carbon steel specimens, after being exposed to corrosion in 0.5 M H_2SO_4, with and without additions of *Buddleia perfoliata,* are shown in Fig. 11. For the uninhibited solution (Fig. 11 a), only a surface showing uniform corrosion can be seen, but after addition of

200 ppm of *Buddleia perfoliata* (Fig. 11 b), a porous, non-protective layer of corrosion products can be seen; however, after addition of 400 ppm of inhibitor (Fig. 11 c), this layer is much more compact, but it still shows some cracks, indicating that it is not protective enough; finally, at the same inhibitor concentration of 400 ppm but at 60 °C (Fig. 11 d), the layer of corrosion products becomes more porous than the one at 25 °C, and becomes less protective, as indicated by all data.

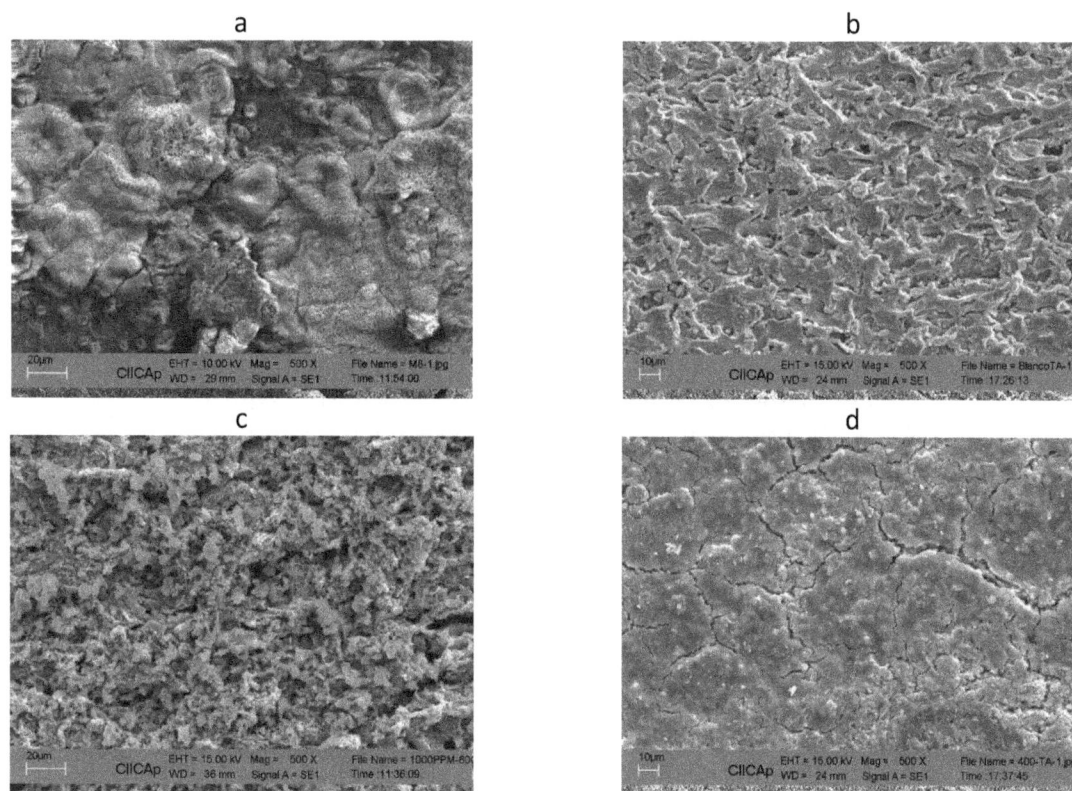

Figure 11. *Microgaphs of 1018 carbon steel corroded in 0.5 M H_2SO_4 with addition of a) 0, b) 200 c) 400 ppm at 25 °C and d) 400 ppm at 60 °C of Buddleia perfoliata*

The use of *Buddleia perfoliata* in traditional medicine has been atributed to the presence of some flavonoids as well as some essential oil, tannic, gallic and oxalic acid [23-24]. UV-visible spectra analysis were performed for the acidic solution containing the extract before and after the corrosion test. For the extract before the corrosion test, the UV-spectrum shows an absorption peak at 360 nm (Fig. 12) corresponding to tannins; after the addition of the extract to the acidic solution, tannins are hydrolyzed producing galic and ellagic acids [25]. Condensed tannins, called pro-antocianidines, are polymers of flavonoids. Formation of a complex formed with Fe^{2+} ions and OH^- groups present in condensed tannins are responsible for the corrosion protection of the metal. As the temperature increases, the degree of polymerization increases and the formed species are more complex and more easily oxidized until the formed corrosion products are detached from the surface and corrosion protection decreases.

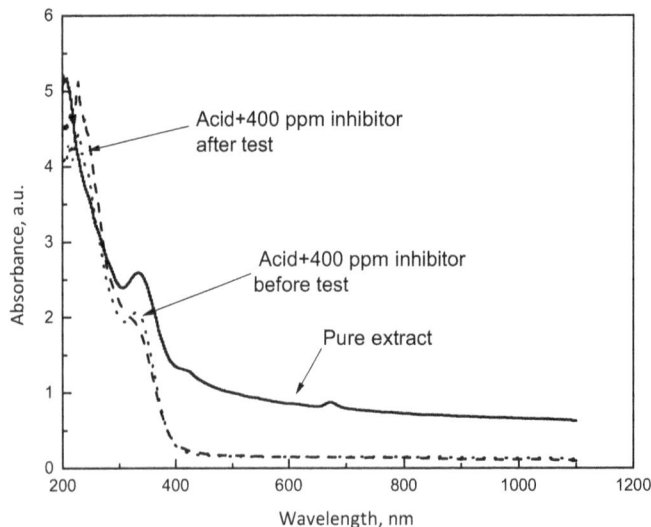

Figure 12. *UV visible spectra of pure Buddleia perfoliata extract,*
0.5 M H₂SO₄ + 400 ppm of inhibitor before and after the corrosion test.

Conclusions

A study of Buddleia *perfoliata* leaves extract as corrosion inhibitor for 1018 carbon steel in 0.5 M H_2SO_4 has been investigated by using electrochemical techniques and weight-loss tests. Results have shown that *Buddleia perfoliata* leaves extract acts as a good inhibitor, and its efficiency increases with increasing the concentration up to 500 ppm but it decreases by increasing the temperature, and remains on the metal surface no more than 12 h. It was found that the inhibitory effect is due to the presence of tannines from this extract which form a protective layer by reacting with Fe^{2+} ions, and which are chemisorbed onto the metal surface following Frumkin type of adsorption isotherm and the corrosion reaction energy barrier is decreased.

References

[1]   H. Ashassi-Sorkhabi, M. Es'haghi, *J. Solid State Electrochem.* **13** (2009) 1297-1301

[2]   M. Behpour, S. M. Ghoreishi, M. Khayatkashani, N. Soltani, *Mater. Chem. Phys.* **131** (2012) 621-633

[3]   L. R. Chauhan, G. Gunasekaran, *Corros. Sci.* **49** (2007) 1143-1161

[4]   P. C. Okafor, M. E. Ikpi, I. E. Uwah, E. E. Ebenso, U. J. Ekpe, S. A. Umoren, *Corros. Sci.* **50** (2008) 2310-2317

[5]   K. F. Khaled, *J. Solid State Electrochem.* **11** (2009) 1743-1749

[6]   M. Z. A. Rafiquee, S. Khan, N. Saxena, M. A. Quraishi, *J. Appl. Electrochem.* **39** (2009) 1409-1417

[7]   H. Ashassi-Sorkhabi, E. Asghari, *J. Appl. Electrochem.* **40** (2009) 631-637

[8]   M. A. Quraishi, A. Singh, V. K. Singh, D. K. Yadav, A. K. Singh, *Mater. Chem. Phys.* **122** (2010) 114-122

[9]   A. Y. El-Etre, *Mater. Chem. Phys.* **108** (2008) 278-282

[10]   A. Bouyanzer, B. Hammouti, L. Majidi, *Mater. Letters* **60** (2006) 2840-2843

[11]   G. O. Avwiri, F. O. Igho, *Mater. Letters* **57** (2003) 3705-3711

[12]   I. B. Obot, N. O. Obi-Egbedi, *J. Appl. Electrochem.* **40** (2010) 1977-1984

[13]   K. W. Tan, M. J. Kassimi, *Corros. Sci.* **53** (2011) 569-574

[14]   P. Lowmunkhong, D. Ungthararak, P. Sutthivaiyakit, *Corros. Sci.* **52** (2010) 30-36
[15]   O. K. Abiola, J. O. E. Otaigbe, O. J. Kio, *Corros. Sci.* **51** (2009) 1879-1881
[16]   K. F. Khaled, *Mater. Chem. Phys.,* **112** (2008) 104 -111
[17]   F. S. De Souza, A. Spinelli, *Corros. Sci.* **51** (2009) 642-649
[18]   C. A. Loto, A. P. I. Popola, *Int. J. Electrochem. Sci.* **6** (2011) 3264-3276
[19]   L. Valek, S. Martinez, *Mater. Letters* **61** (2007) 148-151
[20]   E. E. Oguzie, *Corros. Sci.* **49** (2007) 1527-1539
[21]   F. Bentiss, M. Lagrenee, M. Traisnel, *Corrosion* **56** (2000) 733-742
[22]   A. G. Ocampo in "Flora del Bajio y de Regiones Adyacentes" eds. J. Rzedowski, G. Calderon, Mexico, (2003) 1-7
[23]   P. J. Houghton, *J. Ethnopharm.* **11** (1984) 293-308
[24]   P. J. Houghton, A.Y. Mensah, N. Lessa, L. Yong-Hong, *Photochemistry* **64** (2003) 385-393
[25]   A. Avila, J. Guillermo, A. Romo de Vivar, *Biochem. System. Ecol.* **30** (2002) 1003-1005
[26]   S. Martinez, I. Stern, *Appl. Surf. Sci.* **199** (2002) 83-90
[27]   A. Popova, E. Sokolova, S. Raicheva, M. Christov, *Corros. Sci.* **45** (2003) 33-58
[28]   T. Szauer, A. Brandt, *Electrochim. Acta* **26** (1981) 1253-1256

17

Electrokinetic removal of heavy metals from soil

Puvvadi Venkata Sivapullaiah[✉], Bangalore Sriakantappa Nagendra Prakash* and Belagumba Nagahanumantharao Suma*

Department of Civil Engineering, Indian Institute of Science, Bangalore – 560 012, India
Department of Civil Engineering, UVCE, Jnana Bharathi Campus, Bangalore University, Bangalore -560 087, India

Abstract

Removal of heavy metal ions from soils by electrokinetic treatment has several advantages. The extent of removal, however, is both soil specific and ion specific. The conditions to be maintained have to be established based on laboratory studies. With a view to maximize the removal of metal ions the trends of removal of heavy metal ions such as iron, nickel and cadmium form a natural Indian kaolinitic red earth during different conditions maintained in the electrokinetic extraction process are studied. A laboratory electrokinetic extraction apparatus was assembled for this purpose. Attempts are also made to elucidate the mechanism of removal of the metal ions from soil. The composition of the flushing fluid, voltage and duration of extraction are varied. While dilute acetic acid has been used to neutralize the alkalinity that develops at the cathode, EDTA solution has been used to desorb heavy metals from clay surface. Generally the extent of removal was proportional to the osmotic flow. Nickel and Cadmium are more effectively removed than iron. The percentage removal of Ni is generally proportional to the osmotic flow but shows sensitivity to the pH of the system. There is an optimum voltage for removal of metal ions from soil. The removal of iron was negligible under different conditions studied.

Keywords
Electric; Clays; Metal ions; Removal; Sorption

Introduction

Heavy metal toxicity occurs far more often and people are exposed to toxic metals on a day-to-day basis in our environment. Military activities are one of the primary contributors to metals contaminated soil problems. Military operations such as small arms training, electroplating and metal finishing operations, explosive and propellant manufacturing and use, and the use of lead based paint at military facilities, have resulted in vast tracts of land being contaminated with metals. Whether heavy metal poisoning comes either from these or from cooking utensils, deodorants, pesticides, *etc.*, all have a devastating effect on the human body. Both soils and ground water are often contaminated with heavy metals ions. The existing methods to removal of these metals, particularly from fine grained soils are not often successful or are not cost effective. Electrokinetic remediation has been used to remediate soil and ground water [1].

Electrokinetic remediation of soils offers several advantages over the other remediation methods that are in widespread use today. These include: (i) It is an in-situ process that is 50-90% less expensive than the currently available metals remediation technologies, such as soil washing, pump and treat, and excavation, which are ex-situ methods; (ii) It is extremely effective in fine-grained low permeability soils where other techniques, such as pump and treat, are not feasible. This is due to the fact that the contaminants are transported under charged electrical fields and not hydraulic gradients. (iii) It is suitable for heterogeneous materials such as soils and pressure-driven flushing processes invariably channel the fluid through the largest size pores, leaving the smaller ones nearly untouched. In low-permeability soils, the electro-osmotically driven flow is insensitive to pore size, which allows a rather uniform flow distribution and is applicable equally to coarse and fine-grained soils.

Electrokinetic extraction of contaminants

Electrokinetic remediation is an in-situ process in which an electrical field is created in a soil matrix by applying a low-voltage direct current (DC) to electrodes placed in the soil. As a result of the application of this electric field, heavy metal contaminants may be mobilized, concentrated at the electrodes, and extracted from the soil.

The mass transport processes induced during electrokinetic treatment are primarily electro-osmosis of water, ion-migration and electrophoretic transport of colloids. The rate and efficacy of these processes are often found to depend upon the mineral and chemical composition of the soil and soil water, and the type, age, distribution, and concentration of the contamination. An excellent review of electrokinetic flow processes in soils is given by Yeung [2].

For practical purposes, electro-osmotic fluid volume flow rate is described as analogous to Darcy's law.

$$Q = k_e I_e A \tag{1}$$

Where Q = fluid volume flow rate m^3/s;

        k_e = coefficient of electro-osmotic conductivity, $m^2/V\ s$

        I_e = hydraulic gradient of electro-osmotic flow, m

        A = total cross-sectional area perpendicular to the direction of fluid flow, m^2.

The values of coefficient electroosmotic conductivity of different soils lie in the narrow range of 1×10^{-9} to 1×10^{-8} $m^2/V\ s$. Therefore, an electric field is a much more effective force in driving fluid through fine-grained soils of low hydraulic conductivity than a hydraulic gradient. During an electrokinetic extraction process, the applied DC electric field can thus drive an effective electroosmotic advection of contaminants through the soil and/or inject enhancement agents into the contaminated soil.

Ionic migration or electromigration is the movement of charged chemical species relative to the movement of pore fluid. Anions (negatively charged ions) are moved towards the anode (positive electrode) and cations (positively charged ions) are moved towards the cathode (negative electrode). The alkali metals and alkali earth metals such as Na, K, Cs and Sr, Ca tend to remain ionic under a wide range of pH and redox potential values. Therefore they are expected to electromigrate and be extracted from soil readily unless they become preferentially sorbed onto solid surfaces and clay interstices. Under ideal conditions, the predominant cation and its accompanying anion may be caused to separate efficiently by electromigration only, for which little or no electroosmotic water advection may be necessary. The ionic mobilities of ions in free dilute solutions, *i.e.,* the velocities of the ions under the influence of a unit electric field, are in the range of 1×10^{-8} to 1×10^{-7} m^2/ V s.

Combining the mechanisms of electroosmotic advection and ionic migration results in the applicability of electrokinetic extraction. Past experience with electrochemical treatment of contaminated porous media has shown that the process is most effective when the transported substances are ionic; surface charged or in the form of small micelles with little drags resistance. The effect of ionic migration of anions may be diminished by that of electro osmotic advection. However the direction of electro-osmotic flow may reverse during a prolonged application of a DC electric field across fine-grained soils. The phenomenon of reverse electro-osmotic flow cannot be described by (1) and is not well understood. Most experimental results obtained to date indicate that ionic migration is the dominant.

Factors governing electro-kinetic removal

There are many practical aspects of the technology that needs to be considered carefully before the technology can be successfully implemented in the field.

Soil type

This technology can be successfully applied to clayey to fine sandy soils. It appears that soil type does not pose any significant limitation on the technology. However, contaminant transport rates and efficiencies depend heavily on soil type and environmental variables. Soils of high water content, high degree of saturation, and low activity provide the most favorable conditions for transport of contaminants by electro-osmotic advection and ionic migration. However, soils of high activity, such as Illite, montmorillonite, and impure kaolinite, exhibit high acid/base buffer capacity and require excessive acid and/or enhancement agents to disrobe and solubilize contaminants sorbed on the soil particle surface before they can be transported through the subsurface and removed. Moreover the high sorptive capacity of the clayey soil for contaminant would further aggravate the problem by retarding contaminant transport.

The effects of the soil mineralogy on removal of chromium from soils by electrokinetics were also investigated by [3]. Their results indicate that the presence of carbonate and hematite can adversely impact the process. Values of hydraulic conductivity in different types of soils within a heterogeneous deposit can vary many orders of magnitude. For a contaminated soil deposit containing interlayer of sand and clay, typical values of hydraulic conductivity of these strata are 1×10^{-4} and 1×10^{-8} m / s, respectively. However, the values of electric conductivity of these soils are still within the order of magnitude. As a result the electric field strengths in the different soil layers will be similar when the externally electric potential is applied across the deposit. As the coefficient of electro-osmotic conductivity is insensitive to soil type, the electro--osmotic fluid volume flow rates in different soil layers will thus be very similar.

Complete removal of those metals that possess complex aqueous and electrochemistry and tendency for speciation and forming hydroxide complexes is particularly difficult under variable pH and redox conditions. The technique has been successful to remove >90 % of heavy metals (arsenic, cadmium, cobalt, chromium, copper, mercury, nickel, manganese, molybdenum, lead, antimony, and zinc) from clay, peat and argillaceous sand [4]; to remove spiked lead from kaolinite

[5,6]; to remove 85-95 % of the original concentrations of cadmium, cobalt, nickel, and strontium from laboratory samples prepared from Georgia kaolinite, Na-montmorillonite, and sand-montmorillonite mixture [7]; to remove cadmium from saturated kaolinite [8].

Contaminant type and concentrations

Available experimental data indicate that removal of heavy metals, radio-nuclides, and selected organics by electrokinetics is feasible. Focus is given for the removal of metals in this study. It is anticipated that pollutant, such as PbO, may dissolve and advance through the soil. The process helps in migration of different contaminants in the soil simultaneously. Therefore, the type of contaminant does not pose a significant limitation on the technology, provided the contaminant does not exist in the sorbed phase on the soil particle surface or as precipitates in the soil pore. However a high concentration of ions in the pore fluid will increase the electrical conductivity of soil and thus reduce the efficiency of electro-osmotic fluid flow. More the strength of the electric field applied may have to be reduced to prevent excessive power consumption and heat generation during the process. Nonetheless the concentration of the contaminant does not pose any insurmountable hurdle to the application of the process.

pH gradient across the electrokinetic cell

An electric field causes electrolysis of water and the pH value will drop at the anode and it will increase at the cathode as the produced ions by electrolysis move corresponding to their charge to the electrode of opposite polarity, causing a pH gradient in the soil. When an electric field is applied to wet soil, the soil pH undergoes transient and spatial variation due to decomposition of water, which in turn affect soil surface properties such as cation exchange capacity, ion (cation and anion) adsorption capacity, and magnitude and sign of the ξ potential. Similarly, the electric gradient does not remain constant in time and space due to the changing resistivity and re-distribution of the charges in the soil; and oxidation/reduction state of the soil, contaminants and also the aqueous solution near the electrodes. Rødsand et al. [9] demonstrated the use of acetic acid to depolarize the cathode reaction and an ion-selective membrane to halt the hydroxyl migrating from the cathode into the soil and enhances electrokinetic extraction of lead while the membrane extraction technique does not enhance the technique as expected.

Voltage and current levels

The electric current densities used in most studies are in the order of a few tens of mA /cm^2. Although a high current intensity can generate more acid and increase the rate of transport to facilitate the contaminant removal process, it increases power consumption tremendously, as power consumption is directly proportional to the square of electric current. An electric current density in the range of 1 - 10 A / m^2 has been demonstrated to be the most efficient for the process. However appropriate selection of electric current density and electric field strength depends on the electrochemical properties of the soil to be treated, in particular the electric conductivity. The higher the electric conductivity of the soil, higher will be the required electric current density needed to maintain the required electric field strength. Electric field strength of the order of 50 V / m can be used as an initial estimate for the process. An optimum electric current density or electric field strength should be selected based on soil properties, electrode spacing, and time requirements of the process.

Effluent chemistry and enhancement scheme

Contaminants can exist in different chemical forms in the subsurface depending on environmental conditions. They can exist as solid precipitates, dissolved solutes in the soil pore fluid and/or bonded species on organic matters in the soil. Among these different forms, only dissolved solutes are mobile and removable by electrokinetic extraction and many other remediation technologies. Transformation processes of the contaminant between different chemical forms are contaminant specific, reversible, and dependent on environmental conditions. Nonetheless, most contaminants can be transformed to their dissolved forms. The acidic environment generated at the anode aids in de-sorption and dissolution of metal contaminants from the soil particle surface. However, the basic environment generated at the cathode can hinder the removal of metal contaminants from the soil. In some cases it is necessary to inject reagents into the soil to enhance solubility and transport of metal contaminants. In an acid environment, some metal ions exist as anionic complexes. Both chemical forms are soluble and thus can be extracted from the soil by the process.

It is thus clear that the success of an electrokinetic extraction of contaminants from polluted soils depends on the physical and chemical properties of the soil, the contaminant, and the operating parameters of the extraction system. It is necessary to understand the relative influences of various mechanisms such as osmotic flow induced and the applicability of various enhancement schemes and conditions to be maintained during extraction.

Materials and methods

Soil used

Red earth used in this study was obtained from Indian Institute of Science, campus Bangalore. The soil was collected by open excavation from a depth of one meter from the natural ground. The soil was dried and passed the IS sieve size of 425 μm. The soil so obtained has clay content of 35 %. The clay content consisted predominantly of kaolinitic mineral. Properties of red earth are summarized in Table 1.

Table 1. *Properties of Red Earth*

Property	Red Earth
Specific gravity	2.73
Liquid limit, %	34.9
Plastic limit,%	17.7
Shrinkage limit, %	13.6
Max dry density, g/cm^3	1.79
Optimum moisture content, %	16.23

Preparations of metal ion solutions

The chemicals, cadmium acetate ($(CH_3COO)_2Cd \cdot 2H_2O$), nickel nitrate ($Ni(No_3)_2 \times 6H_2O$), ferric chloride ($FeCl_3$), acetic acid ($CH_3COOH$) solution, hydrochloric acid (HCl), used in this study. All these solution were prepared by adding required quantity of chemicals in deionised water. Standard 100 ppm solutions of cadmium, nickel and iron were prepared by dissolving 2.372 g, 4.955 g and 2.904 g respectively in distilled water and make it to 1000 ml. The solutions were acidified with a drop of HCl. The concentrations of the collected fluid were measured using Atomic Absorption Spectrophotometer after standardization.

Laboratory electrokinetic extraction cell

As the experimental apparatus required for this study program was not readily available in India from conventional geotechnical testing equipment, a new apparatus designed, fabricated and assembled at the Geotechnical Laboratory, Department of Civil Engineering. Indian Institute of Science, Bangalore, India has been used.

The Electrokinetic cell was fabricated using nylon material for the body, which is a non-conductor of electricity, corrosive resistant, not affected by acid or alkali. The electrokinetic cell consists of two end caps and a specimen cylinder made up of nylon. The test sample was 80 mm in diameter and 300 mm long. Ten electrical measurement nodes have been installed on the specimen cylinder at 25 mm intervals so that the electrical voltage distribution along the sample can be monitored continuously during the test. The end caps house the graphite plate electrodes, inflow and outflow tubings. As porous graphite is extremely expensive and fabrication of porous graphite electrodes is very labor intensive, normal grade graphite plates were used in this study. Holes of 1 mm diameter were drilled through the graphite plate electrode to facilitate water transport during electro-osmosis. Details of electrokinetic cell are shown in Fig. 1 The required electrical circuit has been designed, connected and assembled. The circuit is designed to facilitate future automation of the experiment by means of computerized data acquisition system. Provision has been made to pass water through the soil compacted in the cell.

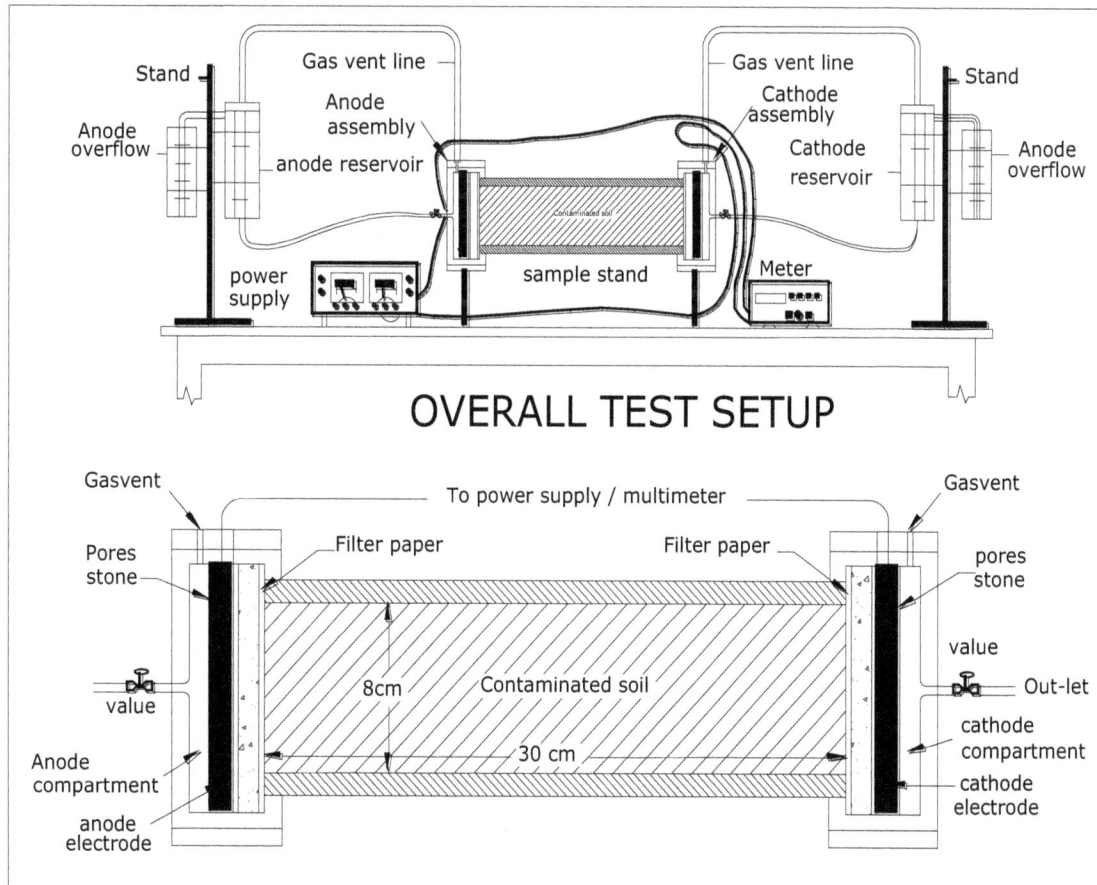

Fig. 1. *Electrokinetic extraction apparatus.*

Electro-Kinetic Extraction

The procedure involves the following sequence,

1. Sample preparation,
2. Variation of flow through red earth with and without contaminants during different phases of extraction,
3. Calculation of electro osmotic permeability, and
4. Estimation of percent removal of metal ions.

The oven dried soil samples were mixed with required amount of water so as to bring it to 110 % of optimum moisture content (OMC) which is about 17.93 %. These samples were kept one day in the desiccators to get uniform moisture content. To prepare contaminated soils with cadmium, nickel and iron solutions of cadmium acetate, nickel nitrate and ferric chloride containing the required amount of chemicals to bring it to 250 mg/kg, 500 mg/kg and 1000 mg/kg of soil mass respectively have been added. The soil samples were then compacted in the cell to bring it to 110 % of maximum dry density on wet of optimum on the compaction curve. The soil is divided into three equal parts by weight and then each part is emplaced into the specimen cylinder and compacted one by one using a screw jack to ensure uniform compaction for the entire specimen.

After compaction, the perforated electrodes were covered with the filter papers and the end caps were closed. Fluid in-let and out let tubes were connected to the cell. Initially, the soil specimen was saturated by maintaining a hydraulic head using self-compensating device and then the head removed. The electrodes were then connected to a constant voltage power supply through the sample and flow is measured at regular intervals. The inlet and outlet were kept at same level such that the hydraulic head difference between them was zero. Then a DC voltage was passed through the cell. The change in the flow, pH and electro osmotic permeability on application of the DC voltage in the soil with and without any contaminant has been studied. After studying the effect of higher voltage the effect of the addition of acetic acid in presence of DC current on the flow, pH and electro osmotic permeability have been studied. Electrokinetic treatment was continued for a specific amount of time. The concentration of the contaminants in the out flow at the cathode induced by electro osmosis was measured using atomic absorption spectrophotometer. The percent removal of contaminants has been calculated.

Results and discussion

Electro-kinetic extraction of Cd, Ni, and Fe (III) ions

The removal of metal ions by induced osmotic flow and ion migration has been studied. The effect of increased voltage on osmotic flow and ion migration has been studied. Attempt has been made to understand the utility of passing acetic acid to control pH and the role of EDTA to desorb and remove contaminant has been studied.

Osmotic flow through soil with and without contaminant

Decontamination of fine grained soil by soil washing is largely inhibited because of their low hydraulic conductivity. One of the important processes by which the decontamination of soils by electro kinetic extraction process is achieved by enhanced flow through fine grained soils on application of voltage across the soil by electro osmosis. Osmotic induces flow through soil even without hydraulic gradient. The extent of osmotic flow through the soil depends on several factors such as type of soil, voltage applied, the type and amount of contaminants present in the soil. The effect of increased voltage across the soil on osmotic flow of fluid through the soil without any contaminants and with known amount of contaminants have been studied

Application of voltage across the soil increases the pH at the cathode and decreases the pH at the anode due to electrolysis of water. Increase in the OH^- ion concentration near the cathode causes high negative value of zeta potential. This in turn increases the flow. For soils containing heavy metal ions the increase in the pH at cathode precipitates ions as hydroxides and reduces the flow rate. The pH at which the ion starts precipitating varies from ion to ion. The effect of this process on the rate of flow through the soil containing different types and levels of contaminant has been monitored.

The efficiency of passage of dilute acetic acid solution during electro kinetic extraction test after noticing the reduction in flow rate in increasing the flow rate has been studied. Increased acidity causes the zeta potential to become less negative. Flow rates have been observed to decrease due to this. However as long as the concentration of ions in the solution is sufficient, electromigration continues even when electro-osmosis has declined. The effect of increase in the voltage while passing of acetic acid to flow rate is monitored.

Application of voltage (5 V) induced flow initially. Increase in voltage (10 V) improves flow by almost 30 times as shown in Fig. 2. Prolonged application with enhanced voltage (20 V) doesn't influence the flow, as it is almost 0. Passing of acetic acid revives flow but the pH remains on the basic front, may be due to deposition of OH^- ions at the cathode. Further increase in voltage (30 V) along with acetic acid compliments the flow marginally due to neutralization of OH^- ions at cathode. Increase in the voltage to 45 V along with acetic acid further enhances the flow considerably and observed to be the maximum during the monitor. However, prolonged application of higher voltage doesn't improve the flow any further, due to increase in pH at the cathode as in Fig. 3.

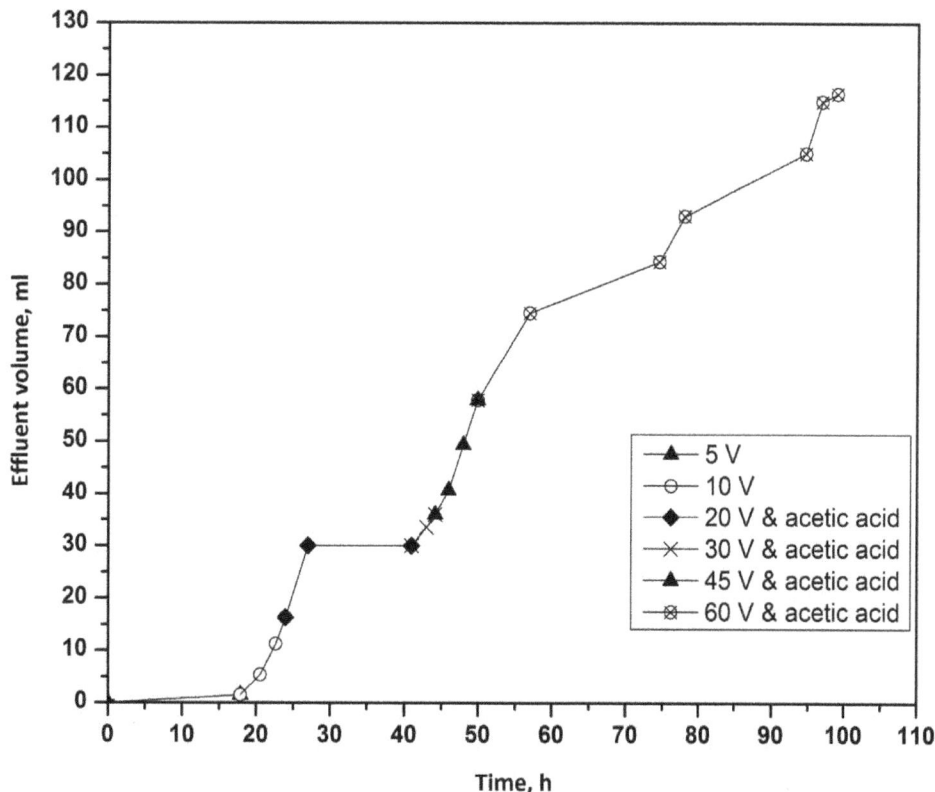

Fig. 2. *Electrokinetic flow through uncontaminated soil*

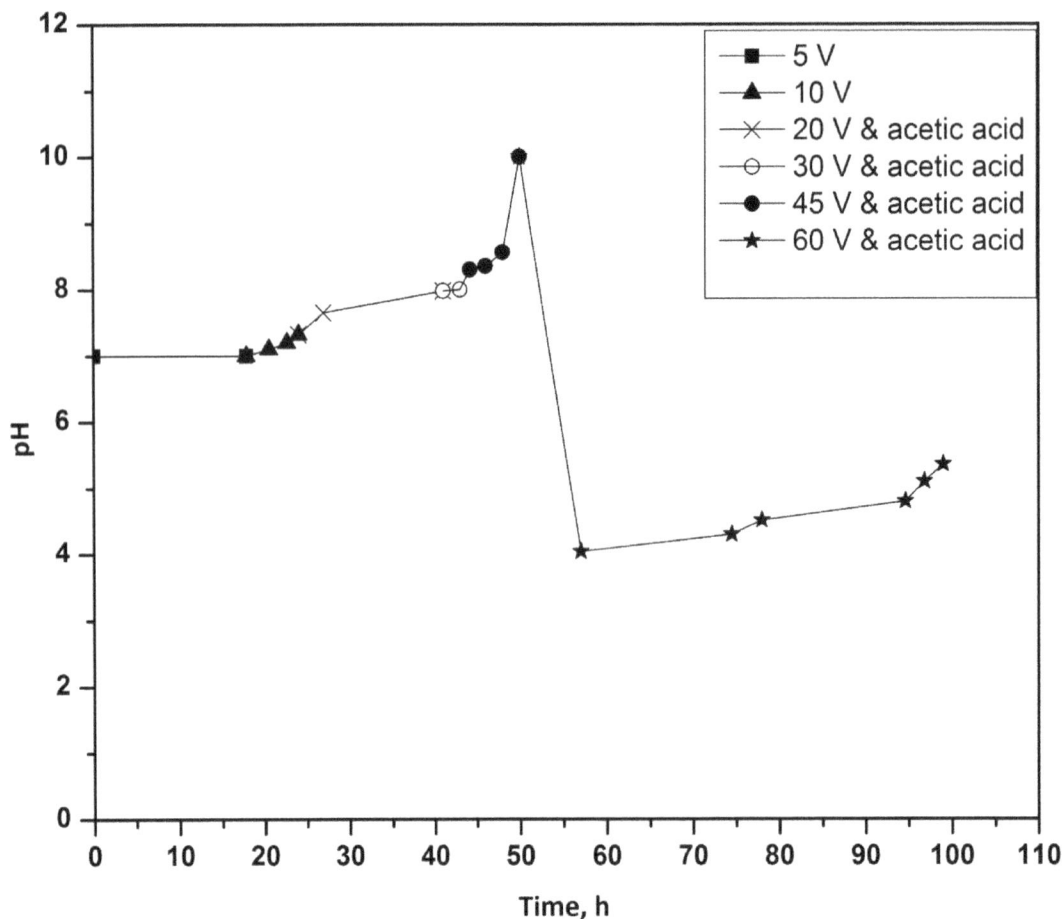

Fig. 3. *Variation of pH of the effluent from uncontaminated soil*

Based on the osmotic flow generated, the osmotic permeability of the soil has been calculated. The electro osmotic permeability of red earth without contaminant is in the range of 2.78 ± 10^{-6} to 4.0×10^{-5} cm^2 / V s as shown in Fig. 4. At 5 V electro osmotic permeability was 2.78×10^{-6} cm^2 / V s. Increase in the voltage to 10V enhances electro osmotic permeability to 4.0×10^{-5} cm^2 / V s being the maximum rate of flow. Increase in voltage to 20 V reduced electro osmotic permeability. Further increase in voltage with acetic acid revived electro osmotic permeability to 1.33×10^{-5} cm^2 / V s. However increase in voltage further (60 V) and prolong application reduced electro osmotic permeability as shown in Fig. 4. Maximum electro osmotic permeability of red earth without contaminant is about 4.0×10^{-5} cm^2 / V s and is observed at 0.33 V / cm.

Effect of nickel ions on the osmotic flow through the soil

To measure the osmotic flow through the soil no hydraulic head was maintained and the flow did not occur. Significant increase in flow was observed on application of initial voltage (5 V) as shown in Fig. 5. Subsequent reduction in flow and increase in pH (Fig. 6), has been attributed to the precipitation of OH$^-$ ions at the cathode. Further enhancement of voltage (10 V) doesn't affect initially but increases the flow on addition of acetic acid at cathode side followed by the sudden reduction in pH. Prolonged application of higher voltages (20, 30, 45 and 60 V) doesn't improve the flow any further, due to increase in pH at the cathode as shown in Fig 5.

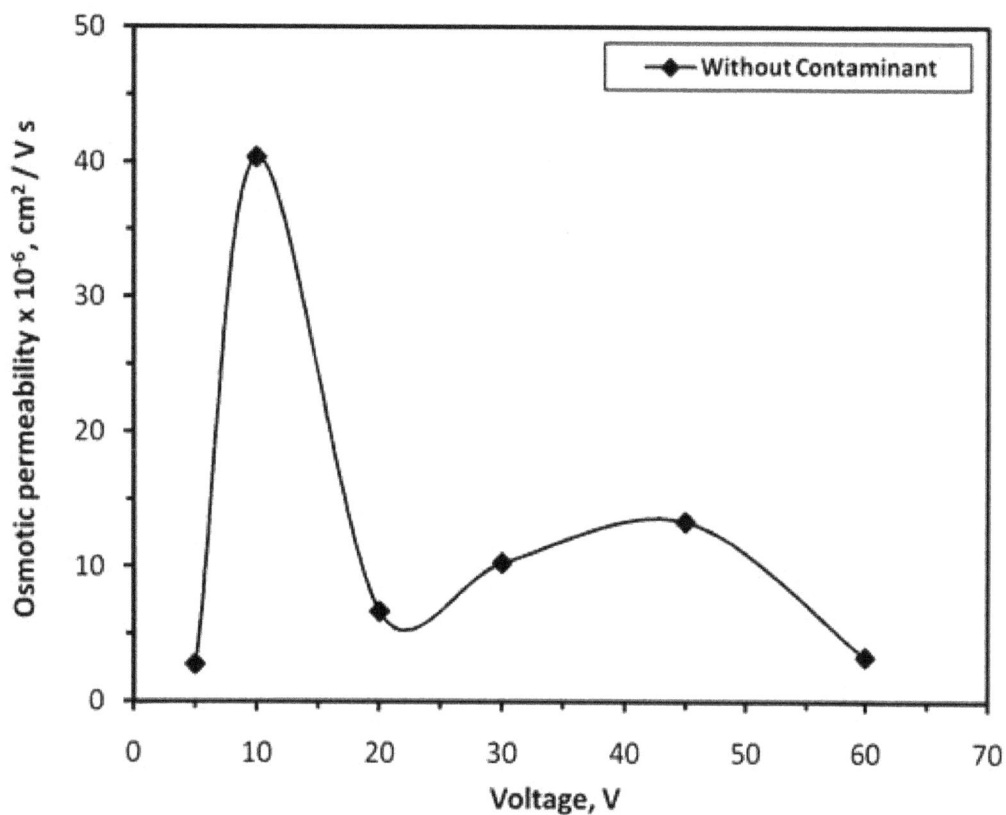

Fig. 4. *Electro-osmotic permeability in uncontaminated soil*

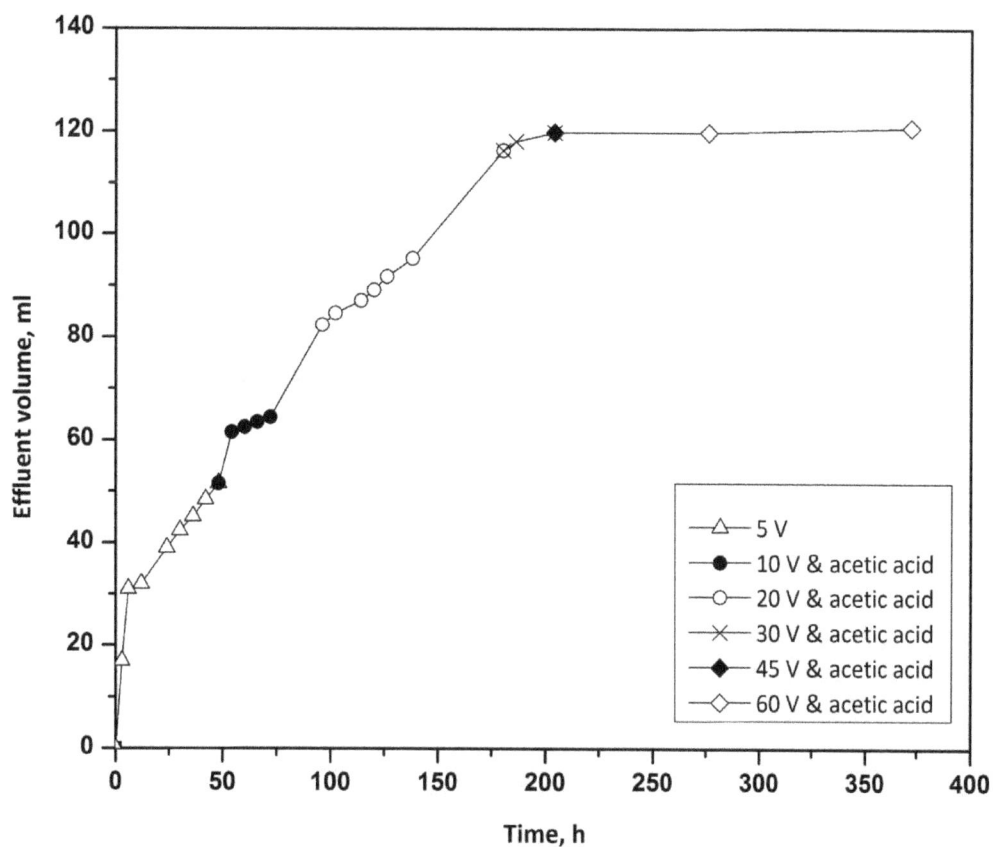

Fig. 5. *Electrokinetic flow through soil contaminated with nickel ions*

The range of electro osmotic permeability with nickel as contaminant in red earth is 0 to 3.55×10^{-5} cm^2/V s as shown in Fig. 7. A maximum electro osmotic permeability of 3.55×10^{-5} cm2 / V s was observed at 5volts. Induction of further voltage (10 V) reduces the electro osmotic permeability by almost 3 times and further increments in the voltage (20, 30, 45, 60 V) gradually reduce the electro osmotic permeability and minimum value (0 cm^2 / V s) at 60 V. The electro osmotic permeability of soil with nickel as contaminant is observed to be maximum value of 3.55×10^{-5} cm^2/ V s, even with a low applied voltage of 0.167 V / cm only.

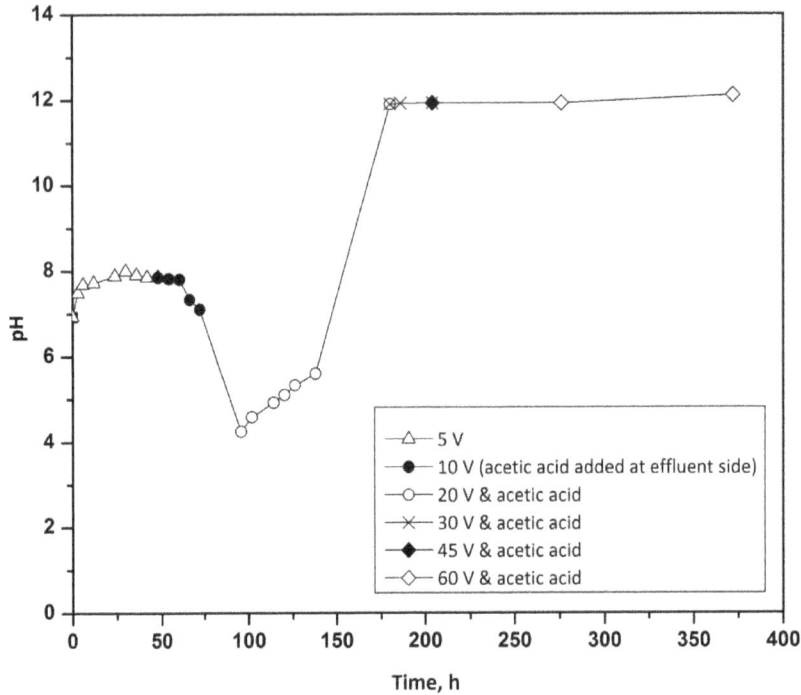

Fig. 6. *Variation of pH of the effluent from soil contaminated with nickel ions*

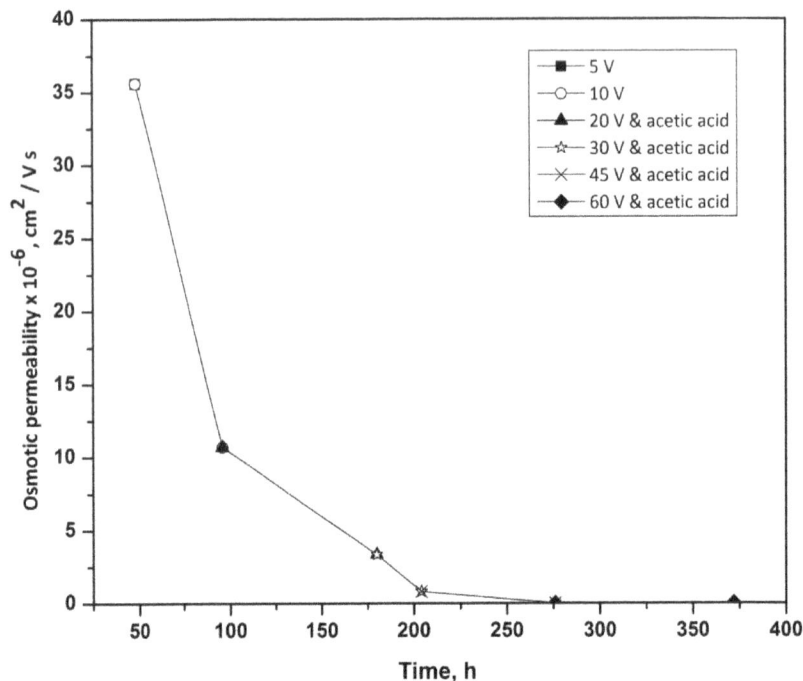

Fig.7. *Electro-osmotic permeability of soil contaminated with nickel ions*

Effect of cadmium ions on the osmotic flow through the soil

Application of voltage (5 V) induced the flow initially with gradual reduction along with time as shown in Fig. 8. Increase of voltage (10 V) doesn't compliment the flow even with addition of acetic acid as shown in Fig. 8, may be due to consequent precipitation of cadmium hydroxide, further increase in voltage doesn't improve the flow, but with addition of acetic acid the flow enhancement was significant. With increase in pH (12.18) the flow reduces in Fig. 9, even with further applications of prolonged voltages.

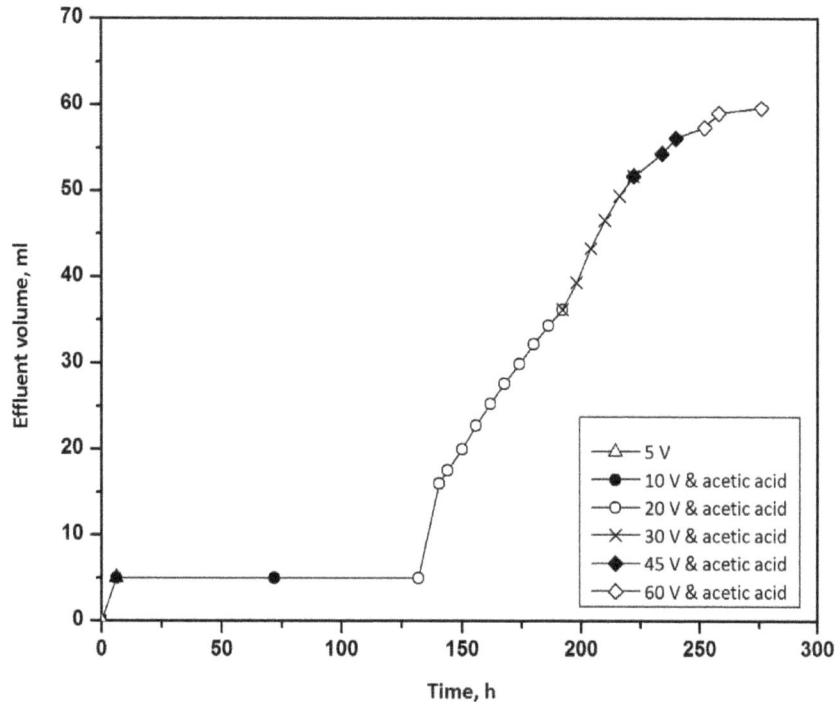

Fig. 8. Electrokinetic flow through soil contaminated with cadmium ions

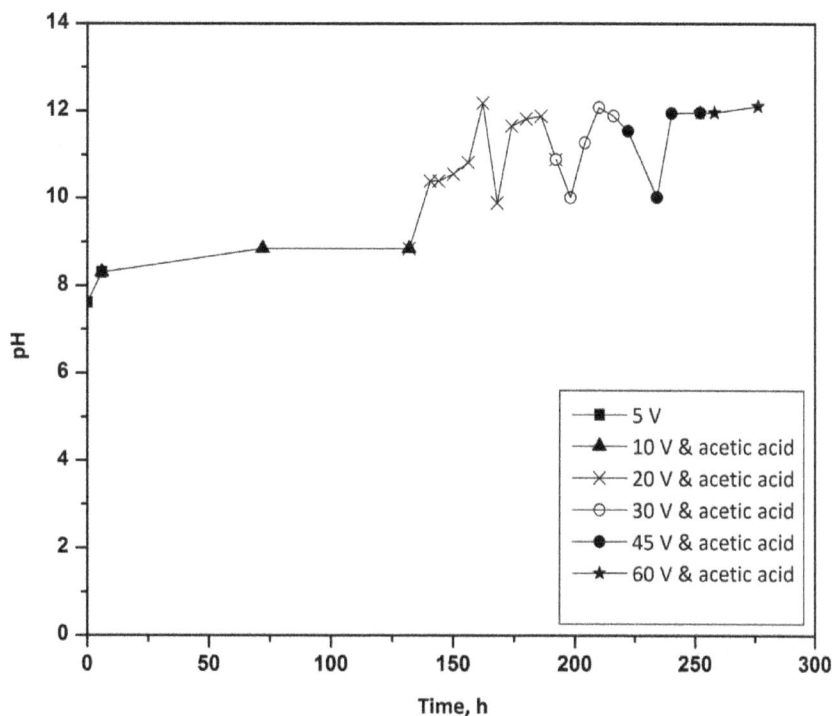

Fig. 9. Variation of pH of the effluent from soil contaminated with cadmium ions

The electro osmotic permeability of red earth with cadmium as contaminant is in the range of 0 to 2.7×10^{-5} cm^2/ V s , Fig. 10. The electro-osmotic permeability was 2.7×10^{-5} cm^2/ V s at a voltage of 5 V with a maximum flow rate. May be due to the precipitation of cadmium hydroxide at cathode the electro osmotic permeability and rate of flow were zero at 10 V.

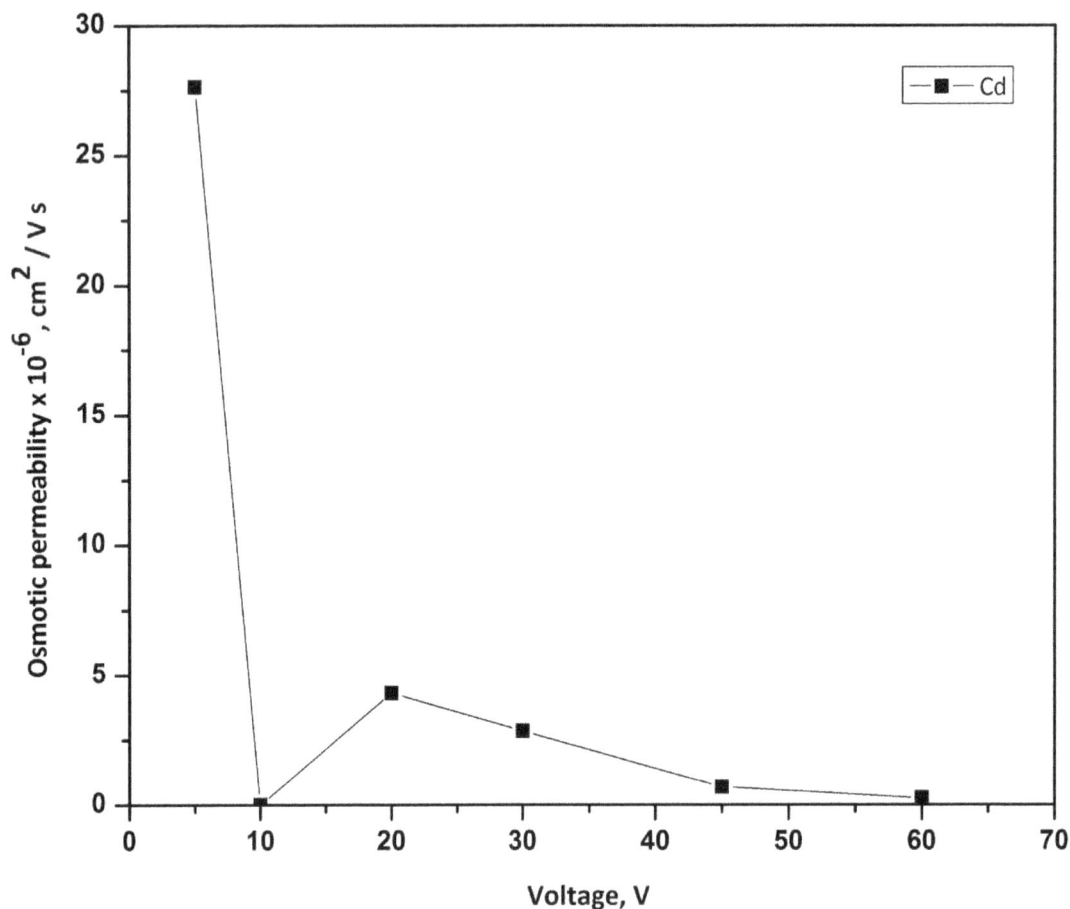

Fig. 10. *Electro-osmotic permeability of soil contaminated with cadmium ions*

A marginal increase was observed with increase in voltage to 20 V. Further increase in voltage with acetic acid resulted in gradual decrease in electro osmotic permeability. Presence of cadmium as contaminant doesn't increase the electro osmotic permeability. Maximum electro osmotic permeability (2.7×10^{-5} cm^2/ V s) is observed at 0.167 V / cm.

Effect of Ferric Ions on the osmotic flow through the Soil

A sudden initial flow was observed on application of 5 volts initially and it subsequent reduction along with time as in Fig. 11. No change was observed when the voltage was increased to 10 V along with addition of acetic acid. Increase in the flow was seen on addition of acetic acid at higher voltage of 20 V, subsequent increase in voltage to 30 V doesn't enhance the flow (Fig. 11), may be due to the low ionic mobility of iron. During the whole process the pH doesn't vary substantially as shown in Fig. 12.

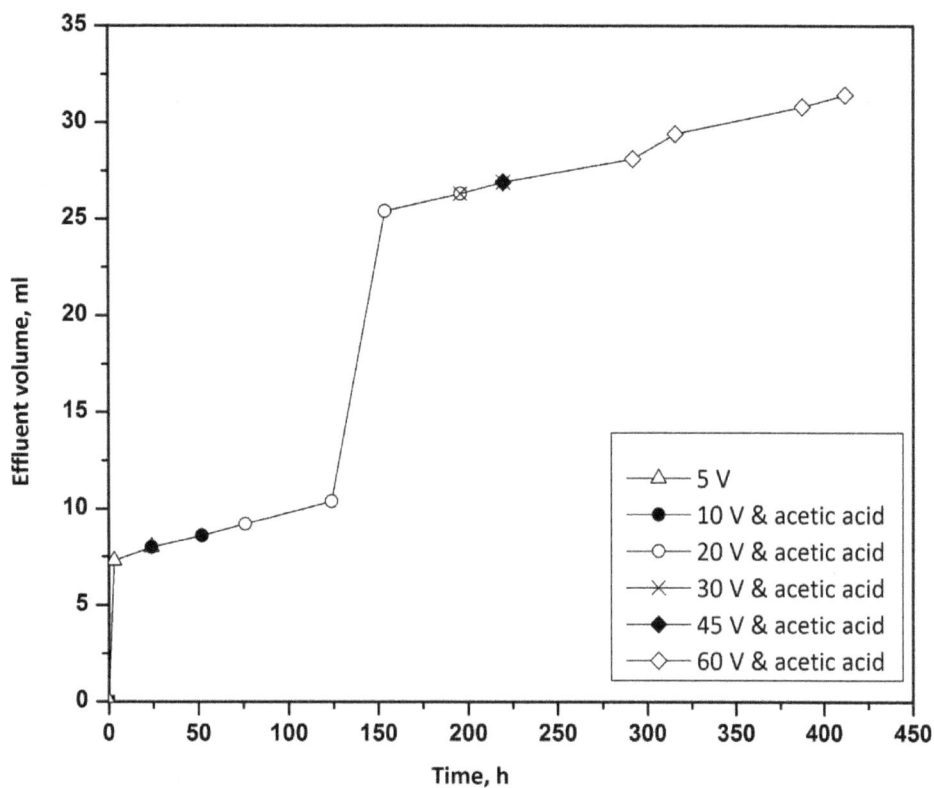

Fig. 11. Electrokinetic flow through soil contaminated with ferric ions

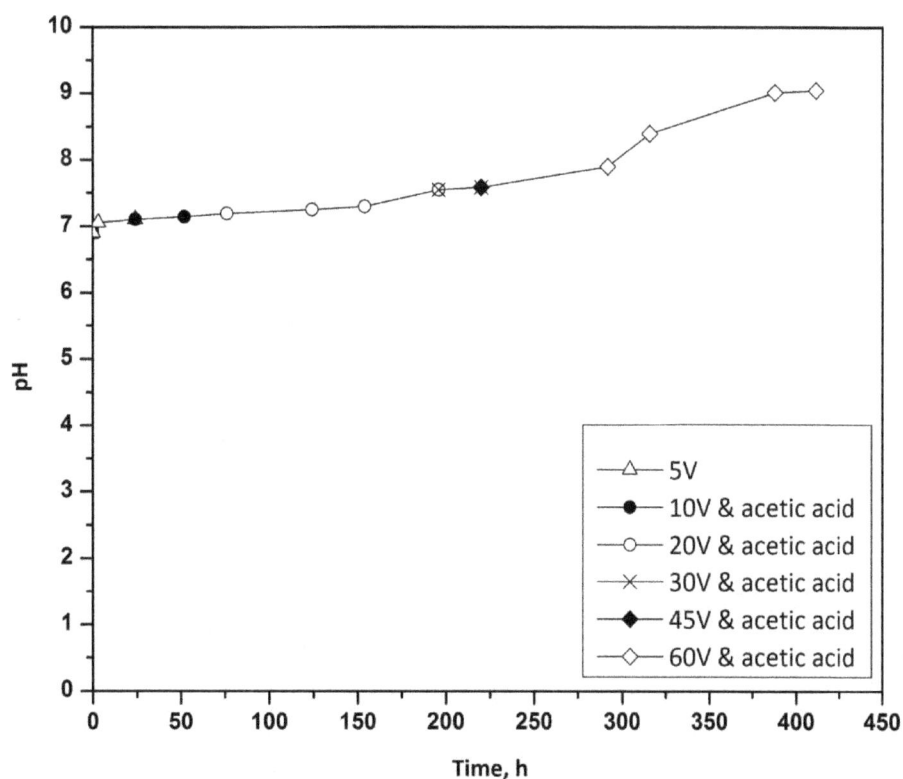

Fig. 12. Variation of pH of the effluent from soil contaminated with ferric ions

The range of electro osmotic permeability of red earth with iron as contaminant is in the range of 6×10^{-7} to 1.1×10^{-5} cm^2 / V s, as shown in Fig. 13. A maximum electro osmotic permeability of 1.1×10^{-5} cm^2 / V s was monitored at 5 volt as shown in Fig. 13. And a sudden decrease of electro osmotic permeability was observed at an applied voltage of 10 V. It subsequently increased with higher voltage of 20 V. The electro osmotic permeability remained at constant rate with further increase in the voltages as shown in Fig. 13. The maximum electro osmotic permeability of 1×10^{-5} cm^2 / V s is observed at 0.167 V / cm.

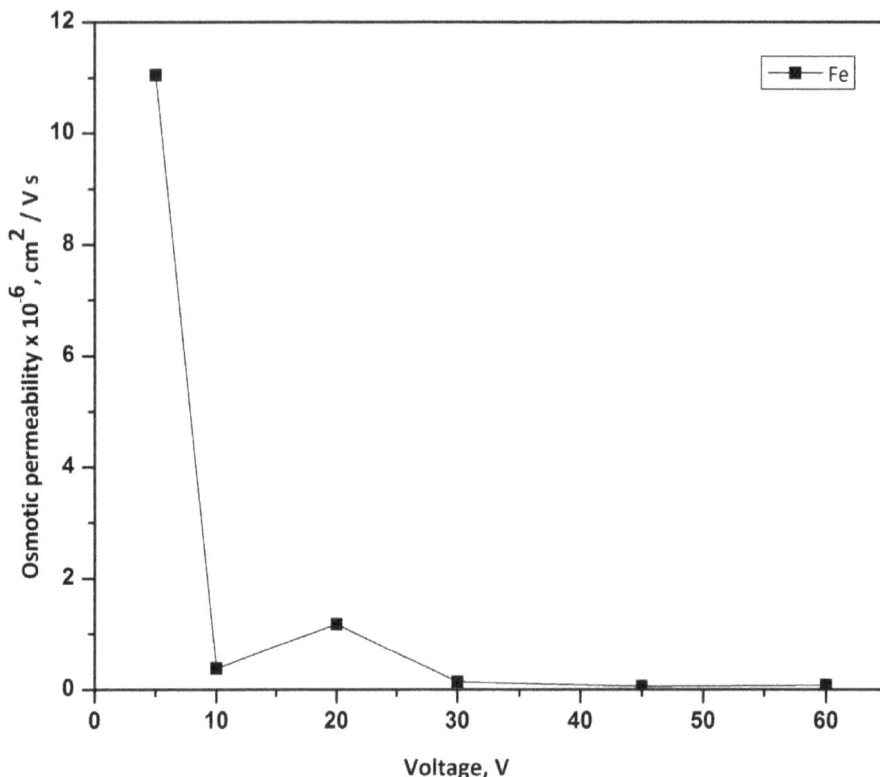

Fig.13. Electro-osmotic permeability of soil contaminated with ferric ions

Electrokinetic extraction of metal ions

The extent of removal of a metal ion from any soil depends on the mobility of the ion and its affinity for the soil clay. The lower is the ionic mobility and the higher is the affinity, the lower is its removal from the soil. The removal also depends upon the solubility product of the species and the pH of the system. As the solubility product at any pH increases their solubility product, the ion is precipitated and hence not removed. The effectiveness of acetic acid to lower the pH and enhance their removal as well as the efficiency of EDTA solutions to enhance desorption of the ion from the soil and hence to increase its removal However during each of these phases the processes might not have been completed. Only the removal trends with different fluids such as water, acetic acid and EDTA solution have been studied. With each fluid the applied voltage is varied. Also the percent removal of the species with respect to increase in osmotic flow has been compared.

How the selected contaminants respond to the removal by electrokinetics with respect to osmotic flow generated at different acidic and alkaline conditions existing during Electrokinetic process are discussed in detail.

Removal of nickel from red earth

It can be seen from Fig. 14 that the amount of removed Ni is almost proportional to the quantity of the flow generated during different phases of electrokinetic process while passing acetic acid and water. This shows that the removal is essentially due to osmotic flow. However, prolonged application of 10 V, while increasing the flow does not proportionally enhance the removal of Ni. This might be due to precipitation of Ni at enhanced pH. Increasing the voltage and passing acetic acid greatly help to remove the Ni contamination. This confirms that Ni is actually precipitated and by passing acetic the rate of removal of Ni is greater than the increased flow due to sudden dissolution of precipitated Ni. Thus passing of acetic acid not only reduces pH at cathode but also enhances the electro-osmotic flow as well as desorption of Ni from the surface of clay particles. With a small increase in voltage from 10 V to 20 V, brings about maximum removal of about 35 %.

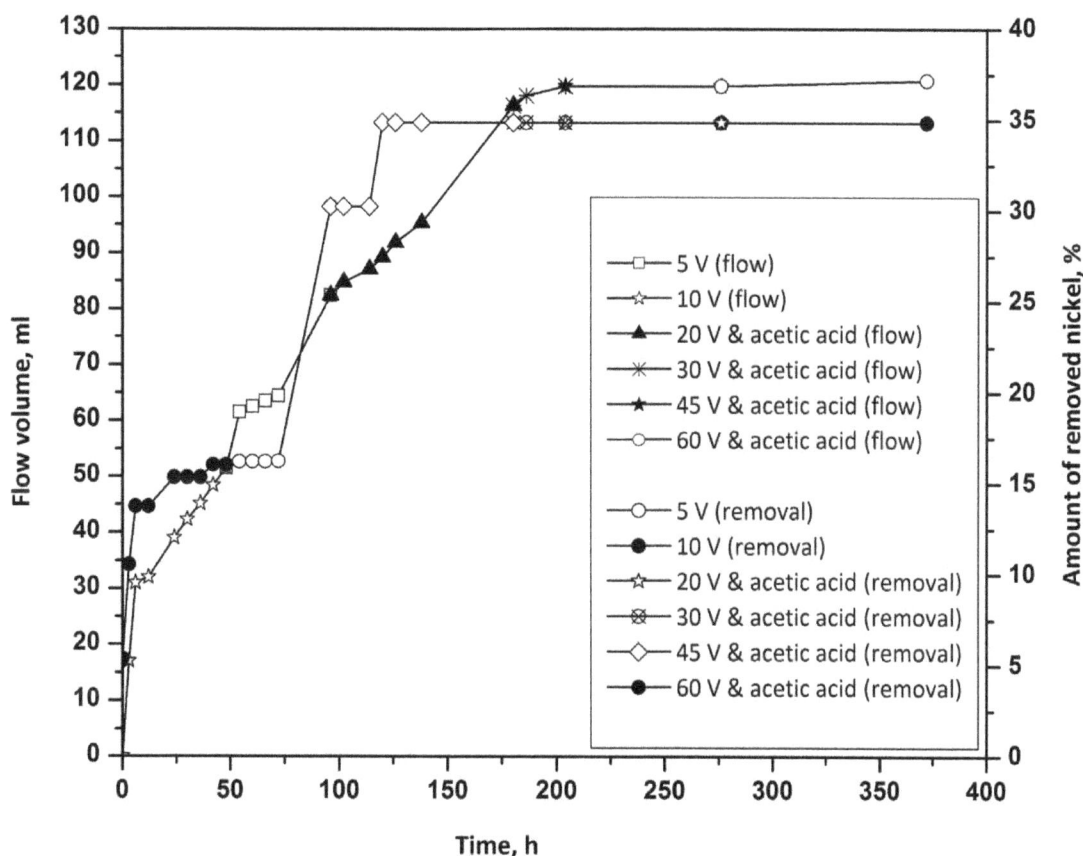

Fig. 14. *Removal trends of nickel ions from soil by electrokinetic process*

As seen from Fig. 14 further increase in voltage (20 - 30 V) increases osmotic flow and amount of removed Cd is also proportional to electro osmotic flow.

Further, prolonged application of voltage beyond 30 V the rate of flow decreases and tends to be more or less constant as result of increasing pH at the cathode. Thus, prolonged application of voltage beyond 30v is not beneficial in extraction of nickel form red earth.

Removal of cadmium from Red Earth

Removal of Cd from red earth is compared with the quantity of flow that is generated at different voltages and by passing water or dilute acetic acid. It can be seen from Fig. 15 that the amount of removed Cd is almost directly proportional to the quantity of flow generated. The variations in the rate of flow at different conditions that exist during the electrokinetic extraction are discussed earlier. There is no phase during which the removal of Cd is greater than the rate of flow. This indicates that either Cd is not precipitated or the precipitated Cd is not dissolved at pH obtained by passing acetic acid. This might be due to the low solubility product of cadmium hydroxide. This indicates that desorbing of Cd from soil is not a major limitation for removal by electrokinetics. During different conditions of removal the maximum removal of Cd is about 16 %. The amount of removed Cd can be enhanced if the rate of flow is maintained for longer periods. The osmotic flow itself is greatly increased by increasing the applied voltage to 20 V along with acetic acid. Increasing the applied voltage beyond 30 V is not beneficial in enhancing the rate flow. Thus, for enhance removal of cadmium passing of about 20 V for longer periods is advisable.

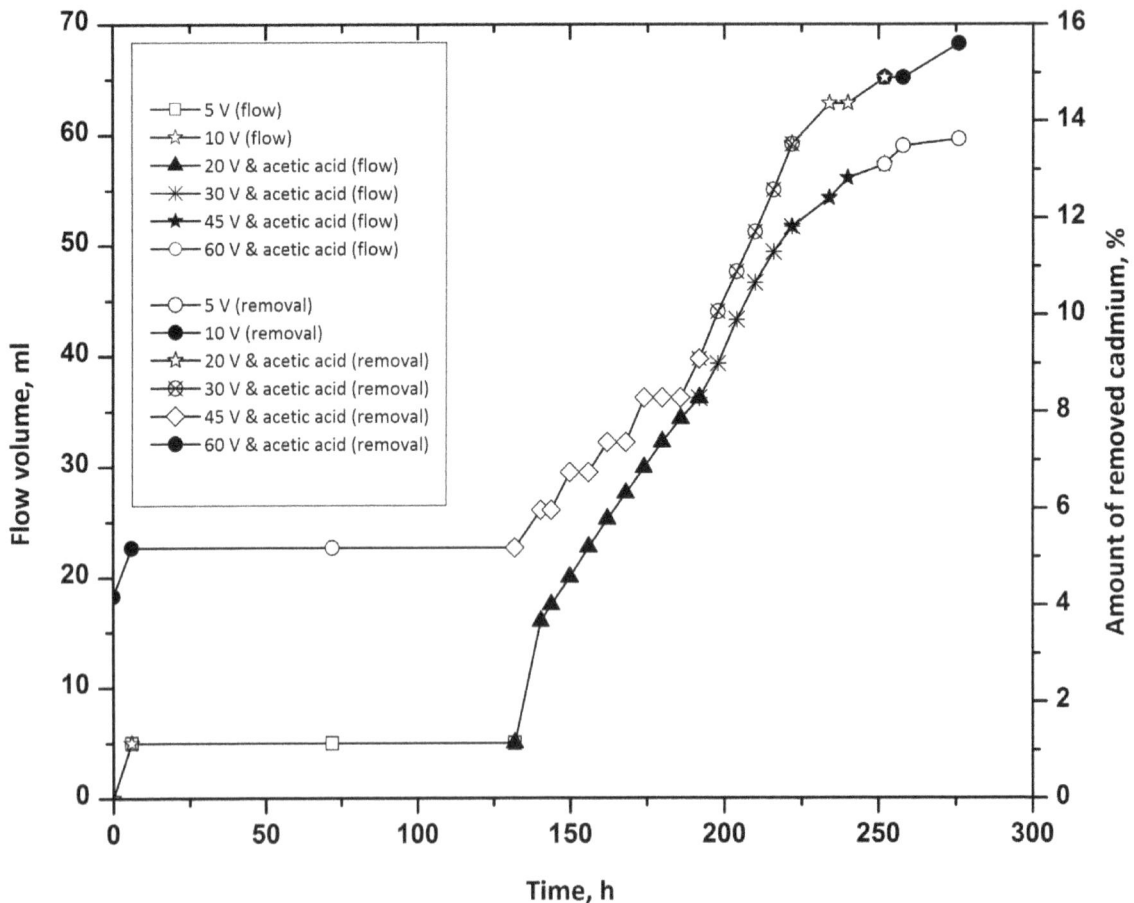

Fig. 15. *Removal trends of cadmium ions from soil by electrokinetic process*

Removal of iron from Red Earth

Percent removal of iron compared to the quantity of flow generated is not proportional at different voltages applied as well as passing of acetic acid solution and water during the electrokinetic extraction process as seen from Fig. 16 due to immediate precipitation as its hydroxide as well as the ion being strongly adsorbed to the clay particle. The solubility product of iron hydroxide is very low and hence is not dissolved and removed. Also, Fe might not have been removed due to its strong adsorption to the clay.

However, with increase in voltage to 60 V *i.e.*, at 2 mV / cm it is observed that enhanced voltage desorbs the ions from clay surface and increase in % removal of the contaminant. Thus, increase in voltage for longer durations might help to enhance its removal. However this method is not very effective for decontamination of iron from soil.

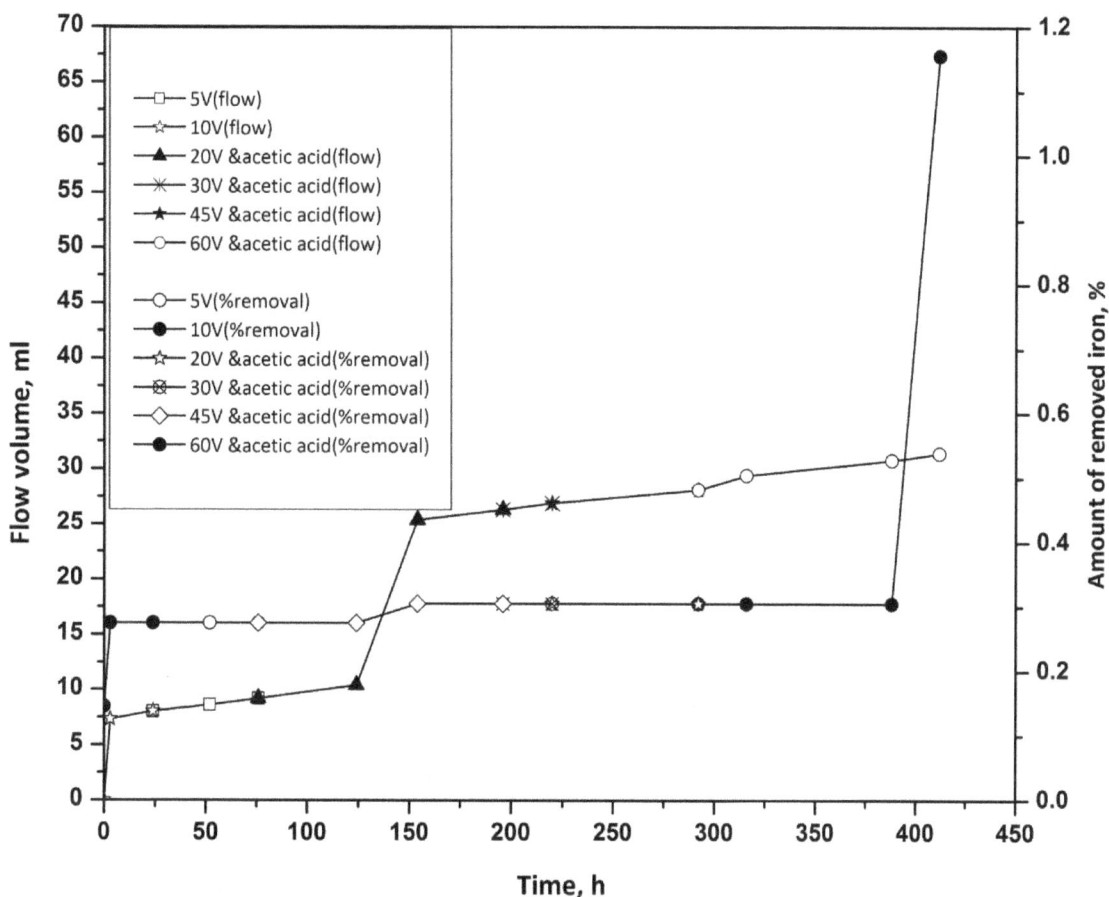

Fig. 16. *Removal trends of ferric ions from soil by electrokinetic process*

Conclusions

1. Osmotic permeability of soil is not significantly affected by the presence of ionic metal contaminants. The variation of osmotic permeability with different contaminants is in the order of **Soil without contaminant > Ni > Cd > Fe.**

2. The rate of flow increases it is not proportional to the applied voltage. Presence of Fe reduces the electro osmotic permeability. The maximum osmotic permeability is obtained 0.167 m V / s. Presence of ionic contaminants reduces the applied voltage required to induce maximum osmotic permeability, though the maximum osmotic permeability induced itself doesn't increase. Passing of acetic acid doesn't enhance the electro osmotic permeability though the pH of the system is controlled.

3. The increase in pH with application of voltage across the contaminated soil increases with Ni and Cd as the contaminants are significantly more.

4. The amount of removed Ni is generally proportional to the osmotic flow but shows more sensitivity to the pH of the system than to the osmotic flow in case of Nickel.

5. The amount of removed Cd contaminant is almost proportional to the osmotic flow generated in case of cadmium.

6. The amount of removed Fe is not proportional to the quantity of osmotic flow generated either by adsorption of the contaminant or if the removed contaminant is precipitated.

7. The higher is the solubility product and lower the affinity of the contaminant the higher is the decontamination of ionic metal contaminants..

References

[1] A. N. Alshawabkeh, Y. B. Acar, *J. Environ. Sci. Heal.* A **27** (1992) 1835-1861.

[2] A. T. Yeung, C. Hsu, R. M. Menon, *J. Geotech. Eng.* ASCE **122** (1996), 666-673.

[3] K. R. Reddy, U. S. Parupudi, S. N. Devulapalli, C. Y. Xu, *J. of Haz. Mat.* **55** (1997) 135-158.

[4] R. Lageman, W. Pool, G. A. Seffinga, *Chem. and Ind.* **18** (1989) 585-590.

[5] Y. B. Acar, A. N. Alshawabkeh, R. J. Gale, *J. Waste Man.* **13** (1993) 141-151.

[6] J. Hamed, Y. B. Acar, R. J. Gale, *J. Geotech. Eng.* ASCE **117** (1991) 241-271.

[7] P. C. Renaud, R. F. Probstein, *J. PhysicoChem. Hydrodyn.* **9** (1987) 345-360.

[8] Y. B. Acar, J. Hamed, A. Alshawabkeh, R. Gale, *Geotechnique ICE*, **44** (1994) 239- 254.

[9] T. Rødsand, B. Acar Yalcin, B. Gijs, *Publikasjon-Norges Geotekniske Institute* **195** (1996) 1518-1535.

Permissions

All chapters in this book were first published in JESE, by International Association of Physical Chemists (IAPC); hereby published with permission under the Creative Commons Attribution License or equivalent. Every chapter published in this book has been scrutinized by our experts. Their significance has been extensively debated. The topics covered herein carry significant findings which will fuel the growth of the discipline. They may even be implemented as practical applications or may be referred to as a beginning point for another development.

The contributors of this book come from diverse backgrounds, making this book a truly international effort. This book will bring forth new frontiers with its revolutionizing research information and detailed analysis of the nascent developments around the world.

We would like to thank all the contributing authors for lending their expertise to make the book truly unique. They have played a crucial role in the development of this book. Without their invaluable contributions this book wouldn't have been possible. They have made vital efforts to compile up to date information on the varied aspects of this subject to make this book a valuable addition to the collection of many professionals and students.

This book was conceptualized with the vision of imparting up-to-date information and advanced data in this field. To ensure the same, a matchless editorial board was set up. Every individual on the board went through rigorous rounds of assessment to prove their worth. After which they invested a large part of their time researching and compiling the most relevant data for our readers.

The editorial board has been involved in producing this book since its inception. They have spent rigorous hours researching and exploring the diverse topics which have resulted in the successful publishing of this book. They have passed on their knowledge of decades through this book. To expedite this challenging task, the publisher supported the team at every step. A small team of assistant editors was also appointed to further simplify the editing procedure and attain best results for the readers.

Apart from the editorial board, the designing team has also invested a significant amount of their time in understanding the subject and creating the most relevant covers. They scrutinized every image to scout for the most suitable representation of the subject and create an appropriate cover for the book.

The publishing team has been an ardent support to the editorial, designing and production team. Their endless efforts to recruit the best for this project, has resulted in the accomplishment of this book. They are a veteran in the field of academics and their pool of knowledge is as vast as their experience in printing. Their expertise and guidance has proved useful at every step. Their uncompromising quality standards have made this book an exceptional effort. Their encouragement from time to time has been an inspiration for everyone.

The publisher and the editorial board hope that this book will prove to be a valuable piece of knowledge for researchers, students, practitioners and scholars across the globe.

List of Contributors

C. SUJAYA
Department of Physics, National Institute of Technology Karnataka, Surathkal, Srinivasnagar
575025, Karnataka, India

H. D. SHASHIKALA
Department of Physics, National Institute of Technology Karnataka, Surathkal, Srinivasnagar
575025, Karnataka, India

G. UMESH
Department of Physics, National Institute of Technology Karnataka, Surathkal, Srinivasnagar
575025, Karnataka, India

A. C. HEGDE
Department of Chemistry, National Institute of Technology Karnataka, Surathkal, Srinivasnagar
575025, Karnataka, India

Mani Nandhini
Department of Chemical Engineering, SSN College of Engineering, Kalavakkam, Chennai 603110, India

Balasubramanian Suchithra
Department of Chemical Engineering, SSN College of Engineering, Kalavakkam, Chennai 603110, India

Ramanujam Saravanathamizhan
Department of Chemical Engineering, SSN College of Engineering, Kalavakkam, Chennai 603110, India

Dhakshinamoorthy Gnana Prakash
Department of Chemical Engineering, SSN College of Engineering, Kalavakkam, Chennai 603110, India

GUOLIANG ZHAN
New Century Concrete of Guangdong Foundation Co., Ltd, Guangzhou 510660, China

JIANCHENG LUO
School of Chemistry and Environment, South China Normal University, Guangzhou 510006, China

SHAOFEI ZHAO
School of Chemistry and Environment, South China Normal University, Guangzhou 510006, China

WEISHAN LI
School of Chemistry and Environment, South China Normal University, Guangzhou 510006, China

Anil N. Ghadge
Department of Civil Engineering, Indian Institute of Technology, Kharagpur -721302, India

Mypati Sreemannarayana
Department of Civil Engineering, Indian Institute of Technology, Kharagpur -721302, India

Narcis Duteanu
University "Politehnica" of Timisoara, Industrial Chemistry and Environmental Engineering, 2 Victoria Sq., 300006 Timisoara, Romania

Makarand M. Ghangrekar
Department of Civil Engineering, Indian Institute of Technology, Kharagpur -721302, India

Sukhdev Singh Bhogal
National Institute of Technical Teachers and Research, Sector-26, Chandigarh, India

Vijay Kumar
Punjab Technical University Campus, SAS Nagar (Mohali)-Punjab, India

Sukhdeep Singh Dhami
National Institute of Technical Teachers and Research, Sector-26, Chandigarh, India

Bahadur Singh Pabla
National Institute of Technical Teachers and Research, Sector-26, Chandigarh, India

OLORUNFEMI MICHAEL AJAYI
Department of Mechanical Engineering, University of Ilorin, Ilorin, Nigeria

JAMIU KOLAWOLE ODUSOTE
Department of Materials and Metallurgical Engineering, University of Ilorin, Ilorin, Nigeria

RAHEEM ABOLORE YAHYA
Department of Materials and Metallurgical Engineering, University of Ilorin, Ilorin, Nigeria

JELENA GULICOVSKI
Vinča Institute of Nuclear Sciences, University of Belgrade, P. O. Box 522, 11000 Belgrade, Serbia

JELENA BAJAT
Faculty of Technology and Metallurgy, University of Belgrade, Karnegijeva 4, 11120 Belgrade, Serbia

VESNA MIŠKOVIĆ-STANKOVIĆ
Faculty of Technology and Metallurgy, University of Belgrade, Karnegijeva 4, 11120 Belgrade, Serbia

BOJAN JOKIĆ
Faculty of Technology and Metallurgy, University of Belgrade, Karnegijeva 4, 11120 Belgrade, Serbia

VLADIMIR PANIĆ
ICTM – Center of Electrochemistry, University of Belgrade, Njegoševa 12, 11000 Belgrade, Serbia

SLOBODAN MILONJIĆ
Vinča Institute of Nuclear Sciences, University of Belgrade, P. O. Box 522, 11000 Belgrade, Serbia

VAISHAKA R. RAO
Electrochemistry Research Laboratory, Department of Chemistry, National Institute of Technology Karnataka, Surathkal, Srinivasnagar-575025, India

A. CHITHARANJAN HEGDE
Electrochemistry Research Laboratory, Department of Chemistry, National Institute of Technology Karnataka, Surathkal, Srinivasnagar-575025, India

K. UDAYA BHAT
Department of Metallurgy and Materials Engineering, National Institute of Technology Karnataka, Surathkal, Srinivasnagar-575025, India

Djamel Ghernaout
Department of Chemical Engineering, University of Blida, PO Box 270, Blida 09000, Algeria
Binladin Research Chair on Quality and Productivity Improvement in the Construction Industry; College of Engineering, University of Hail, PO Box 2440, Ha'il 81441, Saudi Arabia

Abdulaziz Ibraheem Al-Ghonamy
Binladin Research Chair on Quality and Productivity Improvement in the Construction Industry; College of Engineering, University of Hail, PO Box 2440, Ha'il 81441, Saudi Arabia

Mohamed Wahib Naceur
Department of Chemical Engineering, University of Blida, PO Box 270, Blida 09000, Algeria

Noureddine Ait Messaoudene
Binladin Research Chair on Quality and Productivity Improvement in the Construction Industry; College of Engineering, University of Hail, PO Box 2440, Ha'il 81441, Saudi Arabia

Mohamed Aichouni
Binladin Research Chair on Quality and Productivity Improvement in the Construction Industry; College of Engineering, University of Hail, PO Box 2440, Ha'il 81441, Saudi Arabia

Marijana Kraljić Roković
Faculty of Chemical Engineering and Technology, University of Zagreb, Marulićev trg 19, Croatia

Mario Čubrić
Faculty of Chemical Engineering and Technology, University of Zagreb, Marulićev trg 19, Croatia

Ozren Wittine
Faculty of Chemical Engineering and Technology, University of Zagreb, Marulićev trg 19, Croatia

Akshatha Shetty
Department of Civil Engineering, NITK Surathkal-575025, India

Katta Venkataramana
Department of Civil Engineering, NITK Surathkal-575025, India

K. S. Babu Narayan
Department of Civil Engineering, NITK Surathkal-575025, India

Sutanwi Lahiri
Laser and Plasma Technology Division, Bhabha Atomic Research Centre, Mumbai-400085, India

Girish Kumar Sahu
Laser and Plasma Technology Division, Bhabha Atomic Research Centre, Mumbai-400085, India

Biswaranjan Dikshit
Laser and Plasma Technology Division, Bhabha Atomic Research Centre, Mumbai-400085, India

Radhelal Bhardwaj
Laser and Plasma Technology Division, Bhabha Atomic Research Centre, Mumbai-400085, India

Ashwini Dixit
Laser and Plasma Technology Division, Bhabha Atomic Research Centre, Mumbai-400085, India

Ranjna Kalra
Laser and Plasma Technology Division, Bhabha Atomic Research Centre, Mumbai-400085, India

Kiran Thakur
Laser and Plasma Technology Division, Bhabha Atomic Research Centre, Mumbai-400085, India

Kamalesh Dasgupta
Laser and Plasma Technology Division, Bhabha Atomic Research Centre, Mumbai-400085, India

Asoka Kumar Das
Laser and Plasma Technology Division, Bhabha Atomic Research Centre, Mumbai-400085, India

Lalit Mohan Gantayet
Beam Technology Development Group, Bhabha Atomic Research Centre, Mumbai-400085, India

Paul-Cristinel Verestiuc
Faculty of Geography and Geology, Al. I. Cuza University of Iasi, 20 A. Carol I Bd., Iasi, 700505, Romania

Igor Cretescu
Faculty of Chemical Engineering and Environmental Protection, Gheorghe Asachi Technical University of Iasi, 73, D. Mangeron Street, Iasi, 700050, Romania

Oana-Maria Tucaliuc
Faculty of Geography and Geology, Al. I. Cuza University of Iasi, 20 A. Carol I Bd., Iasi, 700505, Romania

Iuliana-Gabriela Breaban
Faculty of Geography and Geology, Al. I. Cuza University of Iasi, 20 A. Carol I Bd., Iasi, 700505, Romania

Gheorghe Nemtoi
Faculty of Chemistry, Al. I. Cuza University of Iasi, 11, Carol I Bd., Iasi, 700506, Romania

Ramakrishnan Kamaraj
CSIR-Central Electrochemical Research Institute, Karaikudi - 630 006, India

Pandian Ganesan
CSIR-Central Electrochemical Research Institute, Karaikudi - 630 006, India

Subramanyan Vasudevan
CSIR-Central Electrochemical Research Institute, Karaikudi - 630 006, India

Abul Hossain
Department of Materials and Metallurgical Engineering, Bangladesh University of Engineering and Technology, Dhaka, Bangladesh

Mohammed Abdul Gafur
Pilot Plant and Process Development Centre (PP & PDC), BCSIR Laboratories, Dhaka, Bangladesh

Fahmida Gulshan
Department of Materials and Metallurgical Engineering, Bangladesh University of Engineering and Technology, Dhaka, Bangladesh

Abu Syed Wais Kurny
Department of Materials and Metallurgical Engineering, Bangladesh University of Engineering and Technology, Dhaka, Bangladesh

ROY LOPES-SESENES
Universidad Autonoma del Estado de Morelos, CIICAP, Av. Universidad 1001, 62209-Cuernavaca, Mor.,Mexico

JOSE GONZALO GONZALEZ-RODRIGUEZ
Universidad Autonoma del Estado de Morelos, CIICAP, Av. Universidad 1001, 62209 Cuernavaca, Mor.,Mexico

GLORIA FRANCISCA DOMINGUEZ-PATIÑO
Universidad Autonoma del Estado de Morelos, Facultad de Ciencias Biologicas, Av. Universidad 1001, 62209-Cuernavaca, Mor., Mexico

ALBERTO MARTINEZ-VILLAFAÑE
Centro de Investigaciones en Materiales Avanzados, Miguel Cervantes 120, Chihuahua, Mexico

Puvvadi Venkata Sivapullaiah
Department of Civil Engineering, Indian Institute of Science, Bangalore – 560 012, India

Bangalore Sriakantappa Nagendra Prakash
Department of Civil Engineering, UVCE, Jnana Bharathi Campus, Bangalore University, Bangalore -560 087, India

Belagumba Nagahanumantharao Suma
Department of Civil Engineering, UVCE, Jnana Bharathi Campus, Bangalore University, Bangalore -560 087, India

www.ingramcontent.com/pod-product-compliance
Lightning Source LLC
Chambersburg PA
CBHW050455200326
41458CB00014B/5187